DOT/FAA/AEE/2012-5
DOT-VNTSC-FAA-12-07

THE ANALYSIS OF MODELING AIRCRAFT NOISE WITH THE NORD2000 NOISE MODEL

Meghan J. Ahearn
Eric R. Boeker
Joyce E. Rosenbaum
Paul J. Gerbi
Christopher J. Roof

U.S. Department of Transportation
Research and Innovative Technology Administration
John A. Volpe National Transportation Systems Center
Environmental Measurement and Modeling Division, RVT-41
Kendall Square
Cambridge, MA 02142

October 2012
Final Report

U.S. Department of Transportation

Federal Aviation Administration

Notice

This document is disseminated under the sponsorship of the Department of Transportation in the interest of information exchange. The United States Government assumes no liability for its contents or use thereof.

Notice

The United States Government does not endorse products or manufacturers. Trade or manufacturers' names appear herein solely because they are considered essential to the objective of this report.

REPORT DOCUMENTATION PAGE		Form Approved OMB No. 0704-0188
	Public reporting burden for this collection of information is estimated to average 1 hour per response, including the time for reviewing instructions, searching existing data sources, gathering and maintaining the data needed, and completing and reviewing the collection of information. Send comments regarding this burden estimate or any other aspect of this collection of information, including suggestions for reducing this burden, to Washington Headquarters Services, Directorate for Information Operations and Reports, 1215 Jefferson Davis Highway, Suite 1204, Arlington, VA 22202-4302, and to the Office of Management and Budget, Paperwork Reduction Project (0704-0188), Washington, DC 20503.	
1. AGENCY USE ONLY (Leave blank)	2. REPORT DATE October 2012	3. REPORT TYPE AND DATES COVERED Final Report
4. TITLE AND SUBTITLE The Analysis of Modeling Aircraft Noise with the Nord2000 Noise Model		5. FUNDING NUMBERS FA4SC3 - LJ200
6. AUTHOR(S) Meghan Ahearn, Eric Boeker, Joyce Rosenbaum, Paul Gerbi, Christopher Roof		
7. PERFORMING ORGANIZATION NAME(S) AND ADDRESS(ES) U.S. Department of Transportation Research and Innovative Technology Administration John A. Volpe National Transportation Systems Center Environmental Measurement and Modeling Division, RVT-41 Cambridge, MA 02142-1093		8. PERFORMING ORGANIZATION REPORT NUMBER DOT-VNTSC-FAA-12-07
9. SPONSORING/MONITORING AGENCY NAME(S) AND ADDRESS(ES) U.S. Department of Transportation Federal Aviation Administration Office of Environment and Energy, AEE-100 Washington, DC 20591		10. SPONSORING/MONITORING AGENCY REPORT NUMBER DOT/FAA/AEE/2012-5
11. SUPPLEMENTARY NOTES FAA Program Managers: Barry Brayer and Keith Lusk (AWP, Western-Pacific Regional Office, Special Programs Staff); Rebecca Cointin and Bill He (AEE, Office of Environment and Energy, Noise Division)		
12a. DISTRIBUTION/AVAILABILITY STATEMENT		12b. DISTRIBUTION CODE
	13. ABSTRACT (Maximum 200 words) This report provides comparisons between AEDT/INM and the Nord 2000 Noise Models for the following parameters: ground type, simple terrain (downward slope, upward slope, hill), temperature and humidity, temperature gradients (positive and negative), turbulence, mixed ground types, hill terrain with mixed ground types, hill terrain with mixed ground types and turbulence, and hill terrain with a positive temperature gradient. The purpose of these comparisons is to highlight portions of the Nord2000 noise propagation methodology that could be considered and adapted for inclusion in AEDT development.	
14. SUBJECT TERMS Aircraft Noise, Noise Prediction, Noise Model Comparison, Nord2000, Aviation Environmental Design Tool, Integrated Noise Model, Ray Model		15. NUMBER OF PAGES 115
		16. PRICE CODE
17. SECURITY CLASSIFICATION OF REPORT Unclassified	18. SECURITY CLASSIFICATION OF THIS PAGE Unclassified	19. SECURITY CLASSIFICATION OF ABSTRACT Unclassified
20. LIMITATION OF ABSTRACT		

NSN 7540-01-280-5500

Standard Form 298 (Rev. 2-89)
Prescribed by ANSI Std. 239-18
298-102

METRIC/ENGLISH CONVERSION FACTORS

ENGLISH TO METRIC

LENGTH (APPROXIMATE)
- 1 inch (in) = 2.5 centimeters (cm)
- 1 foot (ft) = 30 centimeters (cm)
- 1 yard (yd) = 0.9 meter (m)
- 1 mile (mi) = 1.6 kilometers (km)

AREA (APPROXIMATE)
- 1 square inch (sq in, in^2) = 6.5 square centimeters (cm^2)
- 1 square foot (sq ft, ft^2) = 0.09 square meter (m^2)
- 1 square yard (sq yd, yd^2) = 0.8 square meter (m^2)
- 1 square mile (sq mi, mi^2) = 2.6 square kilometers (km^2)
- 1 acre = 0.4 hectare (he) = 4000 square meters (m^2)

MASS – WEIGHT (APPROXIMATE)
- 1 ounce (oz) = 28 grams (gm)
- 1 pound (lb) = 0.45 kilogram (kg)
- 1 short ton = 2000 pounds (lb) = 0.9 tonne (t)

VOLUME (APPROXIMATE)
- 1 teaspoon (tsp) = 5 milliliters (ml)
- 1 tablespoon (tbsp) = 15 milliliters (ml)
- 1 fluid ounce (fl oz) = 30 milliliters (ml)
- 1 cup (c) = 0.24 liter (l)
- 1 pint (pt) = 0.47 liter (l)
- 1 quart (qt) = 0.96 liter (l)
- 1 gallon (gal) = 3.8 liters (l)
- 1 cubic foot (cu ft, ft^3) = 0.03 cubic meter (m^3)
- 1 cubic yard (cu yd, yd^3) = 0.76 cubic meter (m^3)

TEMPERATURE (EXACT)
[(x-32)(5/9)] °F = y °C

METRIC TO ENGLISH

LENGTH (APPROXIMATE)
- 1 millimeter (mm) = 0.04 inch (in)
- 1 centimeter (cm) = 0.4 inch (in)
- 1 meter (m) = 3.3 feet (ft)
- 1 meter (m) = 1.1 yards (yd)
- 1 kilometer (km) = 0.6 mile (mi)

AREA (APPROXIMATE)
- 1 square centimeter (cm^2) = 0.16 square inch (sq in, in^2)
- 1 square meter (m^2) = 1.2 square yards (sq yd, yd^2)
- 1 square kilometer (km^2) = 0.4 square mile (sq mi, mi^2)
- 10000 square meters (m^2) = 1 hectare (ha) = 2.5 acres

MASS – WEIGHT (APPROXIMATE)
- 1 gram (gm) = 0.036 ounce (oz)
- 1 kilogram (kg) = 2.2 pounds (lb)
- 1 tonne (t) = 1000 kilograms (kg) = 1.1 short tons

VOLUME (APPROXIMATE)
- 1 milliliter (ml) = 0.03 fluid ounce (fl oz)
- 1 liter (l) = 2.1 pints (pt)
- 1 liter (l) = 1.06 quarts (qt)
- 1 liter (l) = 0.26 gallon (gal)
- 1 cubic meter (m^3) = 36 cubic feet (cu ft, ft^3)
- 1 cubic meter (m^3) = 1.3 cubic yards (cu yd, yd^3)

TEMPERATURE (EXACT)
[(9/5) y + 32] °C = x °F

QUICK INCH - CENTIMETER LENGTH CONVERSION

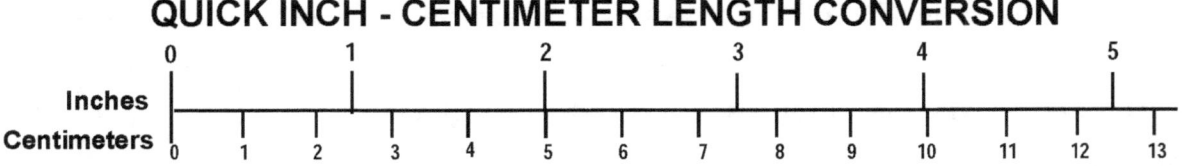

QUICK FAHRENHEIT - CELSIUS TEMPERATURE CONVERSION

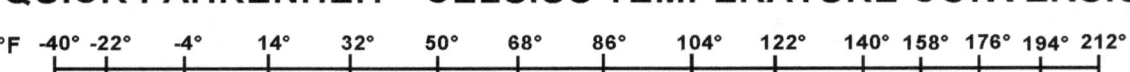

For more exact and or other conversion factors, see NIST Miscellaneous Publication 286, Units of Weights and Measures. SD Catalog No. C13 10286

TABLE OF CONTENTS

Section	Page
EXECUTIVE SUMMARY	1
1 INTRODUCTION	2
2 SOURCE DEFINITION	4
3 ACOUSTIC COMPUTATION METHODOLOGY	5
3.1 INM Acoustic Computation Methodology	5
3.2 Nord2000 Acoustic Computation Methodology	6
4 COMMON STUDY PARAMETERS FOR COMPARISONS	9
5 BASELINE COMPARISON	10
5.1 Baseline Comparison Input Parameters	10
5.2 Baseline Comparison Results	10
6 GROUND EFFECTS	13
6.1 INM vs. Nord2000 Capability	13
6.2 Effective Flow Resistivity (EFR) Comparisons	15
6.2.1 EFR Comparison Input Parameters	15
6.2.2 EFR Comparison Results	15
7 TERRAIN	18
7.1 INM vs. Nord2000 Capability	18
7.2 Terrain Comparisons	19
7.2.1 Terrain Comparison Input Parameters	19
7.2.2 Terrain Comparison Results	19
7.2.3 Terrain Comparison Summary	30
8 WEATHER EFFECTS	32
8.1 INM vs. Nord2000 Capability	32
8.2 Atmospheric Absorption Comparisons	33
8.2.1 Atmospheric Absorption Conditions	33
8.2.2 Atmospheric Absorption Comparison Results	33
8.3 Atmospheric Profile Comparisons	39
8.3.1 Atmospheric Profile Input Parameters	39
8.3.2 Homogeneous Atmosphere Comparison Results	40
8.3.3 Homogeneous Humidity and Temperature Comparison Summary	52
8.3.4 Atmospheric Profile Comparison Results	53
8.3.5 Temperature Gradient Comparison Summary	59

	8.4 Turbulence Comparisons	60
	8.4.1 Turbulence Comparison Input Parameters	60
	8.4.2 Turbulence Comparison Results	60
9	**MULTIPLE PARAMETER VARIATION COMPARISONS**	**63**
	9.1 Multiple Parameter Variation Input Parameters	63
	9.2 Multiple Parameter Variation Results – Mixed Ground Type	64
	9.3 Mixed Ground Type Comparison Summary	73
	9.4 Multiple Parameter Variation Results – Hill Terrain Type	74
	9.5 Hill Terrain, Multiple Parameter Variation Comparison Summary	85
10	**RESULTS SUMMARY**	**86**
11	**CONCLUSIONS AND RECOMMENDATIONS**	**88**
	11.1 Conclusions	88
	11.2 Recommendations	89
	11.2.1 Recommendations for Additional Research	90
Appendix A	**References**	**92**
Appendix B	**Source Model for Nord2000 Analysis**	**94**
Appendix C	**Atmospheric Absorption Data**	**96**
Appendix D	**Noise Model Capability Chart**	**108**

EXECUTIVE SUMMARY

This report presents a review of the Nord2000 outdoor sound propagation prediction method; a curved ray tracing modeling method for computing outdoor sound propagation under a variety of different environmental conditions, such as changes in weather, terrain and ground type. This review includes an evaluation of the noise propagation and environmental effects modeling methodology in Nord2000 and a comparison between the Federal Aviation Administration's (FAA) environmental noise modeling tools (Aviation Environmental Design Tool [AEDT] and Integrated Noise Model [INM]) and a software implementation of Nord2000. A variety of different modeling cases covering a range of environmental conditions were investigated with the purpose of evaluating the benefit of implementing desired Nord2000 software features (whole or in part) into AEDT.

General trends that appear throughout the comparisons described in this report are that (1) the Nord2000 software and INM have the best agreement closest to the source and (2) Nord2000 software computes higher LAMAX levels than INM in most comparisons. The largest deviations from the general INM-Nord2000 software comparison trends occur when a shadow zone is formed, as in the case of the negative temperature gradient and hill terrain comparisons. The effect that shows the largest deviation when compared to the baseline (of the same model) occurs in negative temperature gradient comparison.

From the research described in this report and the confidence in the results yielded, it is recommended to pursue specific adjustments, and not implement the full Nord2000 method into AEDT. Recommendations are made in Section 11.2 to implement ground classes into AEDT and the SAE-ARP 5534 atmospheric absorption standard, once published. Additional research is recommended in Section 11.2 on vertical temperature gradients, pre-defined weather classes, turbulence, and terrain effects with a focus on hill terrain.

1 INTRODUCTION

Nord2000 is a transportation noise modeling methodology that takes into account effects from a variety of different environmental conditions on noise propagation. The Nord2000 methodology was developed as part of a joint Nordic research project led Delta Acoustics & Vibration (Denmark) with a development team consisting of Provinngsanstalt (Sweden), SINTEF (Norway) and VTT (Finland)[*]. Software complying with the Nord2000 methodology was developed by SINTEF for the Norwegian Public Road Administration (NPRA). Avinor A.S. of the Ministry of Transport and Communications of the Kingdom of Norway participated in a software exchange with the FAA, in order to exchange the Nord2000 software (including the Nord2000 Road source model) for the AEDT's Aircraft Performance Module (APM).

In order to avoid confusion between the Nord2000 software implementation developed by SINTEF and the Nord2000 standard, the Nord2000 software implementation will be referred to as Nord2000 and the Nord2000 standard will be referred to as Nord2000 methodology in this report.

The goal of this research is to explore the potential of incorporating some of the Nord2000 acoustic propagation and attenuation methodology into AEDT. The Nord2000 methodology describes a curved ray tracing model that computes outdoor sound propagation under a variety of different environmental conditions including attenuation effects due to weather, turbulence, ground impedance, barrier shielding, and barrier diffraction. Although the implementation of Nord2000 received by FAA through the software exchange with Avinor A.S. is primarily intended for modeling highway, railway and industrial noise, see the Applicability section of the Nord2000 Summary Report[3]. This project explores the Nord2000 implementation's applicability to aircraft noise modeling. In addition, AEDT and INM are primarily empirical, integrated models that represent both acoustic source data and acoustic propagation are represented by the Noise-Power-Distance (NPD) curves in the AEDT database, whereas Nord2000 is a theoretical, ray model that models acoustic propagation and must interface with an external aircraft acoustic source database. Certain Nord2000 adjustments can be calculated independent of the propagation path and could be applied to other models, such as AEDT. The effects investigated in this report

---

[*] The project manager on the Nord2000 development project was Jørgen Kragh of Delta Acoustics & Vibration.

include ground type (effective flow resistivity), terrain, atmospheric absorption, humidity and temperature, vertical temperature gradients, and turbulence. Differences in capability between INM and the Nord2000 methodology are outlined in Appendix D.

This report is organized into 11 sections. In Section 2, source definition is discussed focusing on how each noise propagation model is designed to receive source data, how the two models differ, and what adjustments needed to be made for the comparisons in this analysis. Section 3 describes the acoustic computation methodologies of each model and identifies differences between them. Section 4 describes the parameters common to all comparisons done in this analysis. Section 5 describes the baseline comparison, highlighting the input parameters and model comparisons. Sections 6 through 8 provide background on each propagation effect (ground effects, terrain and weather effects, respectively) followed by a comparison description, comparison input parameters, and observed effects on each model through comparison results. Section 9 describes additional comparisons that vary more than one propagation effect. Section 10 summarizes the results from all comparisons from Sections 5 through 9. Conclusions and recommendations are presented in Section 11. References can be found in Appendix A, and a detailed description of the acoustic source data adjustment used in this analysis is presented in Appendix B. A detailed comparison of atmospheric absorption data is presented in Appendix C. An INM and Nord2000 capability summary chart is provided in Appendix D.

2 SOURCE DEFINITION

The first version of Nord2000 methodology does not specify a noise source model. Instead, the software implementations of Nord2000 often include their own source models, many of which have been developed for road and rail vehicles. In the current Nord2000 Road source model, each source is defined by three different point sources to account for source directivity[3]. Passby events are simulated by a distribution of point sources along the line of travel. Nord2000 is designed to accept sound power levels in third octave bands from 25Hz-10 kHz. However, adjustments can be made to allow for an input of sound pressure level at 1 m from a single omnidirectional source to be accepted, by accounting for the correction factor of $10 \log \frac{1}{4\pi} = -11$ dB between sound power and sound pressure levels[5].

AEDT/INM use Noise-Power-Distance (NPD) and aircraft spectral class data to define a source[1]. NPD data are a set of engine-specific, operation type-specific and metric-specific noise levels, expressed as a function of aircraft engine power setting[*] and distance. NPDs are based on aircraft noise measurement data and take into account noise propagation with spherical spreading, atmospheric absorption and aircraft speed at a range of distances from 200 ft to 25000 ft. The NPD data are then corrected for a variety of aircraft and environmental adjustments. For frequency-based adjustments, the spectral class data are also used. An aircraft spectral class is defined as a one-third octave-band aircraft spectrum, which represents a set of aircraft grouped together based on similar spectral characteristics for similar operational modes. INM uses one-third octave frequency bands from 50 Hz to 10 kHz.

For accurate comparison, an in-house tool was developed to transform the AEDT/INM NPD data into Nord2000 input data. The process takes the spectral class data calibrated to the corresponding NPD at 1000 ft, and corrects that spectrum to a distance of 1m by removing spherical spreading and atmospheric absorption. A detailed description of the acoustic source data adjustment process used in this analysis is presented in Appendix B.

---

[*] Aircraft engine power setting is usually expressed as the corrected net thrust per engine

3 ACOUSTIC COMPUTATION METHODOLOGY

INM and Nord2000 contain fundamentally different acoustic computation methodologies. The sections below describe each methodology including detail on how ground surface, weather, and terrain effects are included in each methodology. Validation information is also presented in each section.

3.1 INM Acoustic Computation Methodology

The core of INM's noise propagation method is the NPD source data. NPDs define a set of noise levels as a function of engine power and distance from the source. The effects of spherical spreading and atmospheric absorption (functions of the propagation distance) are built into the NPDs and automatically accounted for in INM for a given distance through interpolation of NPD data. The slant distance between source and receiver is determined, the closest two distances on the NPD curves are identified, and a log-linear interpolation between the associated NPD level data at those two distances is performed. The interpolated level represents the source level attenuated by spherical spreading and atmospheric absorption at the receiver distance. Adjustments that are applied to the source representation could include the following, based on applicability: ground surface, terrain, weather and aircraft source effects. For a full description of each adjustment refer to reference[1].

Ground surface and terrain effects in INM include ground absorption (as part of lateral attenuation) and line of sight blockage.

Weather effects are modeled as atmospheric absorption, and acoustic impedance adjustments. The atmospheric absorption adjustment is used to correct the reference, SAE-AIR-1845-calculated atmospheric absorption, to airport-specific conditions based on spectral class, temperature, and relative humidity, using the SAE-ARP-866A standard. The acoustic impedance adjustment is also used to correct reference conditions to those representative of a specific airport, based on observer temperature, pressure, and elevation. There is no equivalent to the acoustic impedance adjustment used in Nord2000. However, the acoustic impedance adjustment magnitude is typically small, often less than a few tenths of a dB and, in more extreme cases, less than 1 dB[1].

INM aircraft source effects include a noise fraction adjustment, duration adjustment, and the airplane shielding component of the lateral attenuation adjustment. Airplane shielding models the directivity of sound caused by engine-installation effects, based on aircraft type and engine-mounting location. Since the noise fraction and duration adjustments are only applied for exposure-based metrics and this report will focus on the LAMAX metric, they are not considered further.

INM version 6.1 was validated for over 700 aircraft events within the vicinity of a major commercial airport[6]. The average results show agreement within 3.2 dB of measured data with a standard deviation of 2.0 dB. Other versions of INM were validated in the same study and the results showed an improving trend with each release.

3.2   Nord2000 Acoustic Computation Methodology

Nord2000's noise propagation methodology is based on geometrical ray theory and the theory of diffraction[7]. In contrast to INM, the Nord2000 methodology allows for the calculation of spherical spreading and atmospheric absorption effects independent of the source data. These effects are added later as adjustments to the source input of sound pressure level at 1m. The Nord2000 methodology also applies a different standard in calculating the atmospheric absorption effect. The implications of this are discussed further in Section 8.2. Similar to INM, Nord2000 applies adjustments to the source representation that could include the following based on applicability: ground surface, terrain, and weather effects. Nord2000 does not apply aviation-specific adjustments.

Ground surface effects are based on the effective flow resistivity input provided by the user. The corresponding ground impedance is calculated with the Delany and Bazley model and used to compute the spherical reflection coefficient, which determines the effect of the ground with a Fresnel zone approach[7]. In contrast, INM always assumes a soft ground, for jet aircraft, and does not provide for varied, user-defined ground types. The effect of ground impedance is investigated in Sections 5 and 9.

In the Nord2000 methodology, terrain effects are determined from the input of a cross-section of the terrain height along the propagation path from source to receiver. To increase computational efficiency, the terrain profile is further simplified into a smaller number of straight line segments. Calculation of the effect of diffraction around terrain features, for example a wedge-shaped screen as shown in Figure 1, can be a function of the distances between source and wedge peak R_S and wedge peak and receiver R_R, the diffraction angles θ_S and θ_R, the wedge angle β, and the reflection coefficients on either side of the wedge Q_S and Q_R[7]. In contrast, the geometry of the INM line-of-sight blockage adjustment is based only on the difference between the direct path length and the length of the path diffracted around the terrain feature. The effect of terrain is further investigated in Sections 7 and 9.

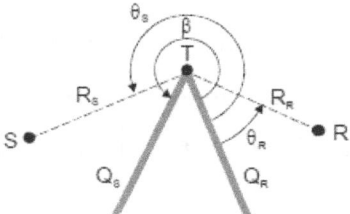

Figure 1. Wedge-Shaped Screen and Associated Parameters Applied in Diffraction Effect Calculations[7]

Weather conditions in the Nord2000 methodology are modeled by an approximate sound speed profile, which varies with height. First, the Nord2000 methodology considers an approximate combined logarithmic-linear profile that captures the effect of both a linear temperature contribution to the sound speed, and a logarithmic wind contribution. Then, the sound speed profile is simplified further by approximating the logarithmic sound speed profile by an "equivalent" linear profile. A curved ray scheme is used to propagate the sound[8]. The effect of turbulence is also included in the Nord2000 methodology through user-defined turbulence strength parameters of wind and temperature, used in the calculation of an incoherence coefficient[7]. There are no equivalent user inputs in INM to specify refractive atmosphere or turbulence parameters. The effects of refractive atmospheres are further investigated in Sections 8.3 and 9, and the effect of turbulence is explored in Sections 8.4 and 9.

Although implemented, scattering effects due to irregularities of urban areas or thick vegetation have not been validated in Nord2000 and therefore have not been investigated in this research.

The Nord2000 methodology has been validated for propagation distances up to 1000 m with measured data, benchmark calculations, and reference results[9]. Validation cases conducted from 0 to 400 m included terrain and weather variation. Validation cases with propagation distances between 400 and 1000 m were only validated for flat terrain and weather variation. The model was originally intended for use over short distances. The stated design goal for an implementation of the Nord2000 methodology[3] was to obtain good accuracy up to 1000 m and acceptable accuracy[*] up to 3000 m.

Individual environmental effects, such as ground and weather effects, are described and analyzed further in Sections 5 through 9. In each section, the specific environmental effect is described and the observed effect on each model is presented through comparison results.

---

[*] Definitions of good and acceptable were not provided.

4 COMMON STUDY PARAMETERS FOR COMPARISONS

A variety of different modeling scenarios covering a range of environmental conditions were compared in this research. The Nord2000 methodology implementation used for the comparisons in this report was the Nord2000 Fortran source code version 16. INM version 7.0b was selected for the AEDT/INM portion of these comparisons because of the maturity of the application.

All comparisons described in this report have study parameters in common. A Boeing 747-400 was modeled in an over-flight operation mode at a source height of 1000 ft. The level over-flight operation mode was chosen due to the consistency of the flight segments and in turn the ability to easily isolate propagation parameters. Eleven equally spaced receivers were setup reaching up to 3000 m laterally from the flight path at a height of 1.219 m. The noise metric calculated was LAMAX. LAMAX was chosen to focus on a segment by segment comparison between the two models, and to isolate relevant noise adjustments. For the baseline comparison, two rows of receivers 8 nmi apart were modeled to show the consistency of the over-flight operation; the remaining comparisons were only done for receiver row 14. The receiver positions relative to the flight track are shown in Figure 2.

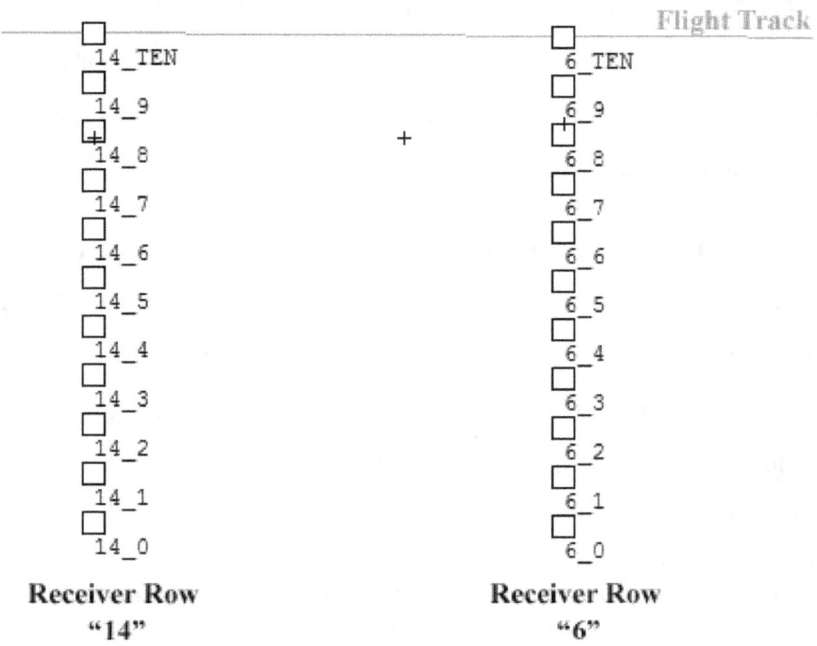

Figure 2. Receiver Position

5 BASELINE COMPARISON

The baseline comparison was run as a reference point for the rest of the comparisons throughout this report. The baseline comparison was run with all INM and Nord2000 settings as close to equal as possible to put differences due to propagation effects into perspective. Default settings were used for parameters that could not be equalized.

5.1 Baseline Comparison Input Parameters

The baseline input parameters are described in Table 1.

Table 1. Baseline Comparison Input Parameters

	INM	Nord2000
Temperature	15°C	15°C
Temperature Gradient	0	0
Relative Humidity	70%	70%
Turbulence	N/A	0
Terrain	Flat	Flat
Ground Type	Soft, 80 CGS Rayls	Soft, 80 CGS Rayls

5.2 Baseline Comparison Results

The results for receiver rows 6 and 14 are shown in Table 2 and Table 3[*]. The results for receiver row 14 are also shown graphically in Figure 3. The LAMAX results are presented as the mean difference +/- the standard deviations of differences between INM and Nord2000 at the 11 receivers. The terrain and ground type are shown in Figure 4.

In the baseline case, Nord2000 computes LAMAX 5.3 +/- 2.6 dB greater than INM under baseline conditions. The maximum difference between the two models is 9.0 dB at the farthest receptor. The differences seen between Nord2000 and INM with baseline conditions are due in part to the differing atmospheric absorption standards used in the model. Nord2000 uses a more detailed atmospheric absorption model, which results in higher LAMAX values. It is expected for differences to increase with distance because most propagation models are designed for short

---

[*] As expected, the two receiver row's output match well for a given noise model; from this point forward comparisons are only presented for receiver row 14. The small differences between the two receiver rows are likely due to the way INM and Nord2000 calculate maximum noise metrics. In INM, the maximum noise level is calculated at each end of the segment and at the closest point of approach. The maximum noise level for the segment is then considered to be the maximum value of those three values, see Section 3.3.2 of the INM Technical Manual[1] for more information. This level of detail is not described for Nord2000 acoustic computation by segment, however it is not expected that Nord2000 follows the same decision making process as INM.

distance propagation (under 7620 m) and have not been validated against measured data at longer distances. Atmospheric absorption is an attenuation in dB per distance, therefore differences between standards are more pronounced with increased propagation distance. Additionally, at close distances, meteorological effects such as wind speed and temperature can be simplified and modeled relatively easily. However, at greater distances the importance and variation of meteorological effects increases and is difficult to model accurately.

Table 2. Baseline, Receiver Row 14 Results

Grid Fourteen	Distance (m)	INM (dB): Baseline	Nord2000 (dB): Baseline	Difference (dB)
14_10	0	79.8	81.8	-1.93
14_9	300	76.0	77.9	-1.8
14_8	600	69.8	72.8	-3.0
14_7	900	65.2	68.8	-3.6
14_6	1200	61.5	65.6	-4.1
14_5	1500	58.0	63.1	-5.1
14_4	1800	55.0	61.1	-6.1
14_3	2100	52.4	59.5	-7.1
14_2	2400	50.2	58.0	-7.9
14_1	2700	48.1	56.6	-8.5
14_0	3000	46.3	55.3	-9.0

Table 3. Baseline, Receiver Row 6 Results

Grid Six	Distance (m)	INM (dB): Baseline	Nord2000 (dB): Baseline	Difference (dB)
6_10	0	79.8	82.0	-2.12
6_9	300	76.0	78.0	-2.0
6_8	600	69.8	72.8	-3.0
6_7	900	65.2	68.8	-3.7
6_6	1200	61.5	65.6	-4.1
6_5	1500	58.0	63.1	-5.1
6_4	1800	55.0	61.1	-6.1
6_3	2100	52.4	59.5	-7.1
6_2	2400	50.1	58.0	-7.9
6_1	2700	48.1	56.6	-8.5
6_0	3000	46.3	55.3	-9.0

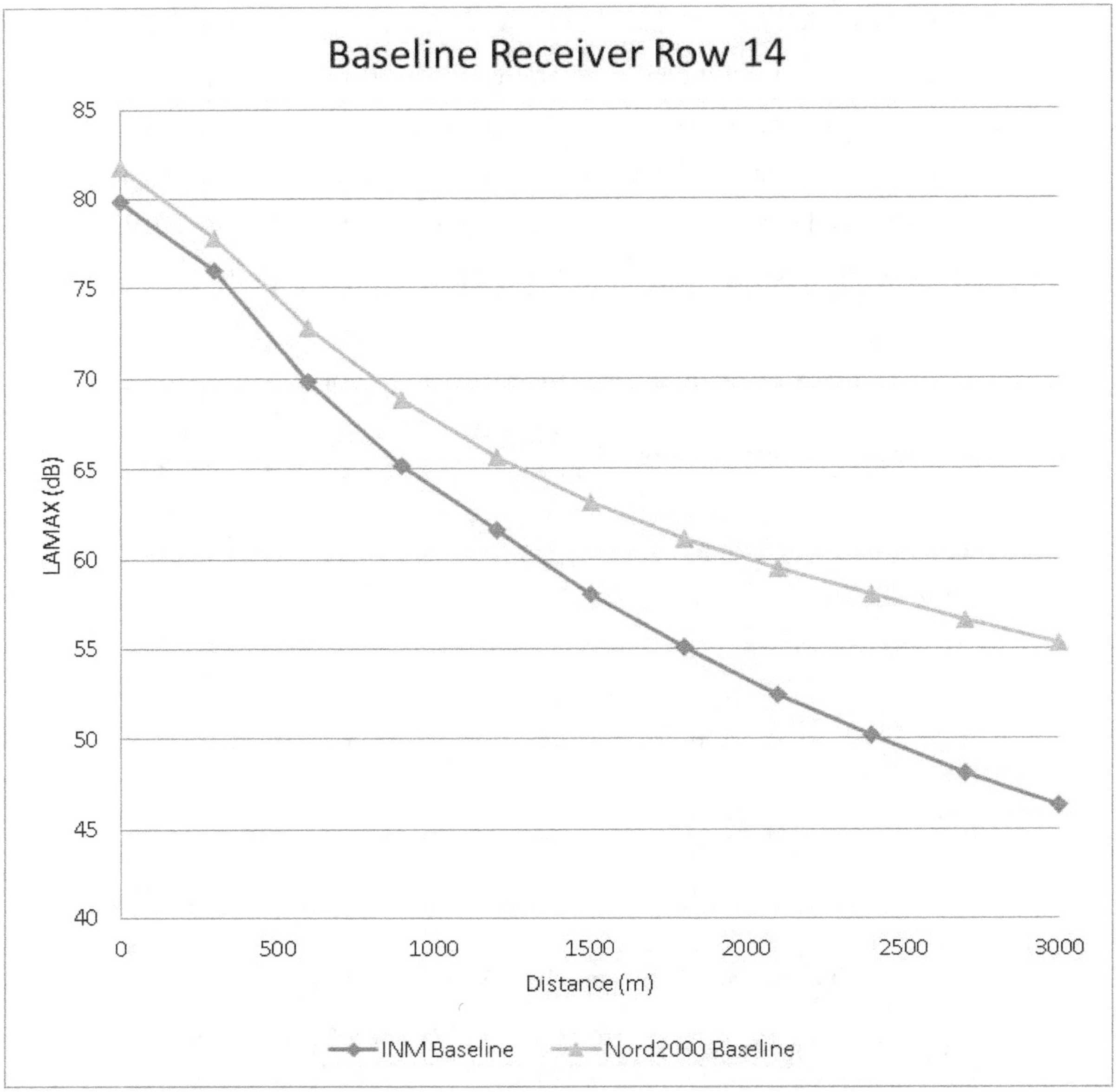

Baseline Comparison   **The Analysis of Modeling Aircraft Noise**
                                                        with the Nord2000 Noise Model

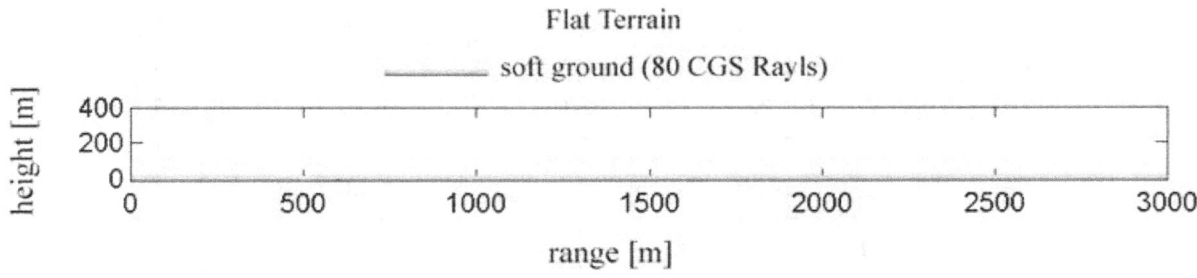

Figure 3. Baseline Results

Figure 4. Baseline Terrain and Ground Type

6 GROUND EFFECTS

6.1 INM vs. Nord2000 Capability

In INM, the lateral attenuation adjustment accounts for ground effect. The adjustment is defined as the difference in sound level directly under the aircraft's flight path and a location offset from the flight path at the time of closest approach. This adjustment was derived from field measurements that were taken over grass-covered terrain[1, 13]. For this research, an effective flow resistivity of 80 CGS Rayls was chosen to approximate the INM soft ground type[*]. It is important to note that the lateral attenuation adjustment includes ground and refraction effects, as well as airplane shielding effects due to aircraft engine installation locations. Under the conditions used in this report, the ground effect component is calculated as

$$G(l_{seg}) = \begin{cases} 11.83 \cdot [1 - e^{-0.00274 \cdot l_{seg}}], & 0 < l_{seg} < 914 \text{ m (3000 ft)} \\ 10.86, & l_{seg} > 914 \text{ m (3000 ft)} \end{cases} \quad \text{Eq. 6-1}$$

where l_{seg} is the sideline distance in the horizontal plane from the source to receiver. The ground effect fits into the full lateral attenuation adjustment calculation as

$$LA_{ADJ(INM)} = -[E_{ENGINE}(\varphi) - \frac{G(l_{seg}) \cdot \Lambda(\beta)}{10.86}] \quad \text{Eq. 6-2}$$

where $E_{ENGINE}(\varphi)$ is the engine installation effect and $\Lambda(\beta)$ is the refractive-scattering component[1].

While INM does account for ground absorption due to reflections off soft ground, it does not separate the contributions of the direct and reflected rays.

---

[*] The effective flow resistivity of 80 CGS Rayls for soft ground was specified in the Nord2000 Road User's Guide[4]. AEDT and INM assume propagation over soft ground through the implementation of a lateral attenuation adjustment to the NPD data, as specified in SAE-AIR-5662[13], however this adjustment is derived from empirical data measured over grass-covered, acoustically soft ground, and does not take into account effective flow resistivity directly in its calculation.

In the Nord2000 methodology, any value can be entered for effective flow resistivity[4]. Table 4 is provided to the user for suggested input to Nord2000. Ground impedance is calculated from the effective flow resistivity using the Delany and Bazley model

$$Z = 1 + 9.08 \left(\frac{1000 f}{\sigma}\right)^{-0.75} + j11.9 \left(\frac{1000 f}{\sigma}\right)^{-0.73} \qquad \text{Eq. 6-3}$$

assuming an $e^{-j\omega t}$ time dependence, where Z is the normalized acoustic impedance of the ground, f is frequency, and σ is the effective flow resistivity in N·s/m^4 (1 CGS Rayl = 1 kN·s/m^4). Ground impedance is used to calculate the reflection coefficients, which determine the reflected ray contribution to the total sound level[7].

The existing impedance class categorization could provide a starting point for implementation of a drop-down selection menu in AEDT/INM by any of the fields in the chart if desired.

Table 4. Classification of ground type[4]

Impedance Class	Representative Flow Resistivity [CGS Rayls]	Description
A	12.5	Very soft (snow or moss-like)
B	31.5	Soft forest floor (short, dense heather-like or thick moss)
C	80	Uncompacted, loose ground (turf, grass, loose soil)
D	200	Normal uncompacted ground (forest floors, pasture field)
E	500	Compacted field and gravel (compacted lawns, park area)
F	2000	Compacted dense ground (gravel road, parking lot, ISO 10844 asphalt)
G	20000	Hard surface (most common asphalt)
H	200000	Very hard and dense surface (dense asphalt, concrete, water)

6.2 Effective Flow Resistivity (EFR) Comparisons

To capture the most extreme difference in the way INM and the Nord2000 methodology model ground effect, a very hard ground type (20000 CGS Rayls) was modeled in the Nord2000 software implementation and compared to the INM baseline condition (since INM only offers a soft ground type for modeling noise from jet aircraft).

6.2.1 EFR Comparison Input Parameters

Nord2000 EFR input parameters are described in Table 5. Parameters that have been changed from the baseline conditions are identified by italicized text.

Table 5. EFR Comparison Input Parameters

	INM –Baseline	Nord2000
Temperature	15°C	15°C
Temperature Gradient	0	0
Relative Humidity	70%	70%
Turbulence	N/A	0
Terrain	Flat	Flat
Ground Type	Soft, 80 CGS Rayls	*Hard, 20000 CGS Rayls*

6.2.2 EFR Comparison Results

The results for the EFR case in Nord2000 compared to the INM baseline (with soft ground, approximately 80 CGS Rayls) are shown in Table 6 and Figure 5. The LAMAX results are presented as the mean difference +/- the standard deviations of differences between INM and Nord2000 at the 11 receivers. The terrain and ground types are shown in Figure 6. Table 7 displays the difference between the Nord2000 hard ground and baseline results.

Nord2000 with hard ground computes LAMAX 7.0 +/- 2.4 dB greater than INM with the baseline conditions. The maximum difference between the two models for this comparison is 10.9 dB, at the farthest receptor. Under hard ground conditions, Nord2000 predicts levels up to 2.0 dB larger than in its corresponding baseline case with soft ground because the ground-reflected rays are attenuated by the ground absorption. Therefore, the mean differences between the INM baseline results and Nord2000 results with hard ground are increased.

Table 6. EFR Results

Receptor Row Fourteen	Distance (m)	INM (dB): Baseline	Nord2000 (dB): Hard Ground	Difference (dB)
14_10	0	79.8	83.7	-3.8
14_9	300	76.0	79.8	-3.8
14_8	600	69.8	74.7	-4.9
14_7	900	65.2	70.9	-5.7
14_6	1200	61.5	67.6	-6.0
14_5	1500	58.0	64.9	-6.9
14_4	1800	55.0	62.6	-7.6
14_3	2100	52.4	60.7	-8.2
14_2	2400	50.2	59.2	-9.0
14_1	2700	48.1	58.1	-10.0
14_0	3000	46.3	57.3	-10.9

Table 7. EFR Comparison, Nord2000 Difference from Baseline

Distance (m)	Nord2000 (dB): Baseline	Nord2000 (dB): Hard Ground	Difference (dB)
0	81.8	83.7	-1.9
300	77.9	79.8	-1.9
600	72.8	74.7	-1.9
900	68.8	70.9	-2.0
1200	65.6	67.6	-1.9
1500	63.1	64.9	-1.8
1800	61.1	62.6	-1.5
2100	59.5	60.7	-1.2
2400	58.0	59.2	-1.2
2700	56.6	58.1	-1.5
3000	55.3	57.3	-2.0

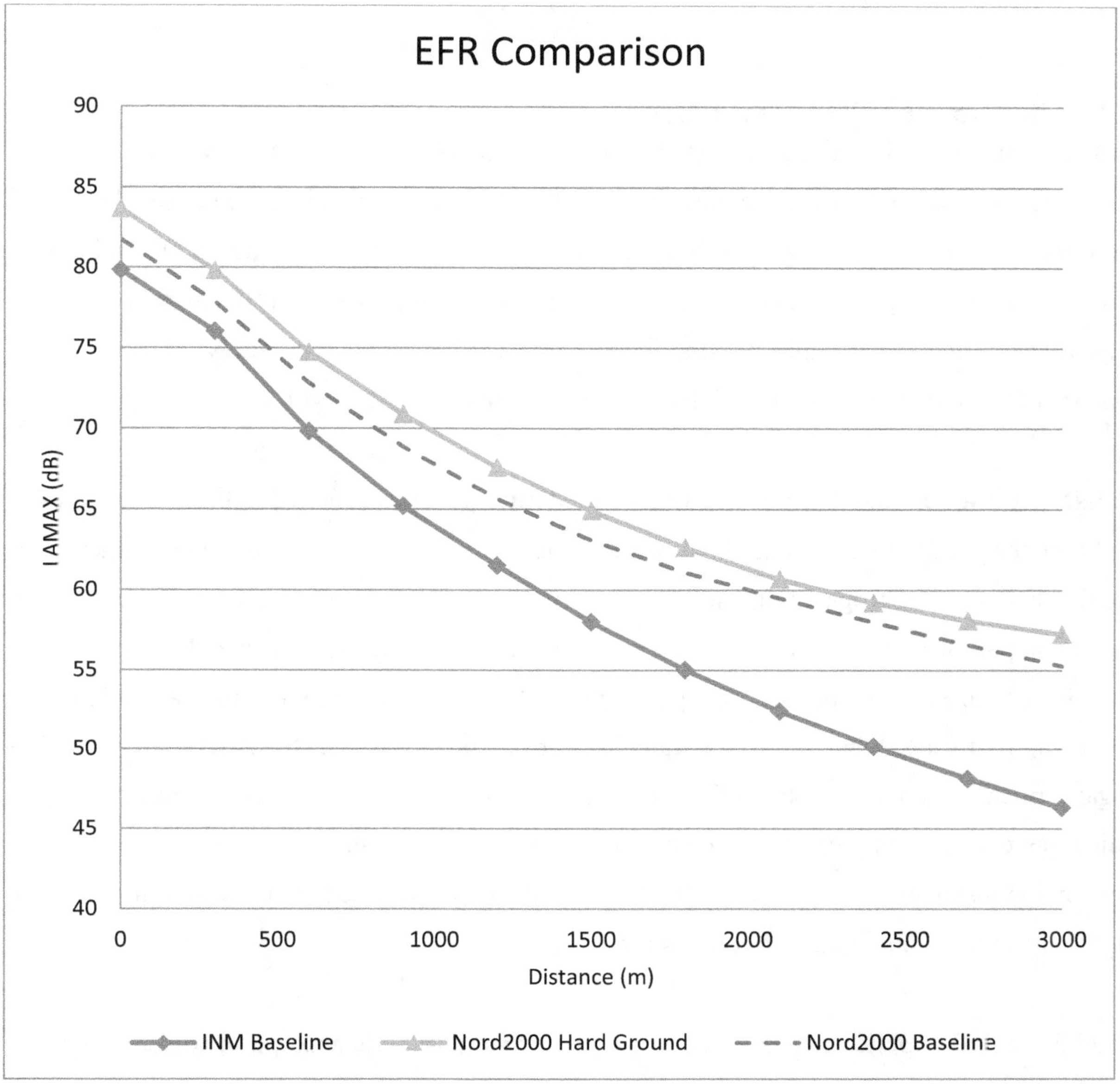

EFR Comparison

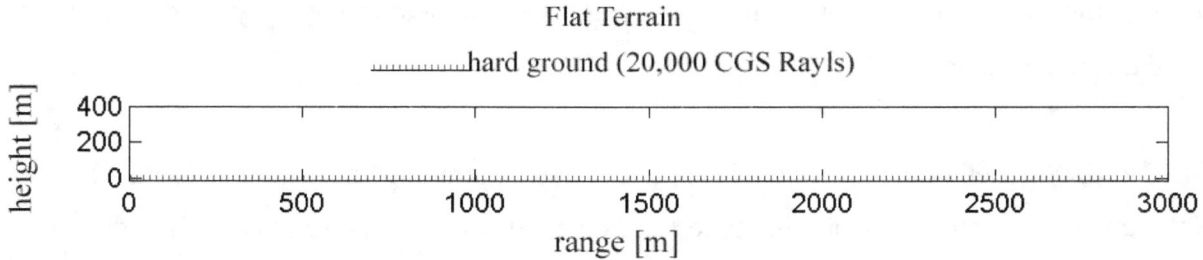

Figure 5. EFR Results

Figure 6. Nord2000 ERF Comparison Terrain and Ground Type

7   TERRAIN

7.1   INM vs. Nord2000 Capability

INM accepts terrain input in a variety of different formats (3CD, National Elevation Dataset (NED) GridFloat, and Digital Elevation Model (DEM)), all of which include terrain elevation information as a function of geographic location[1]. Custom terrain files with 10 m spacing were generated by forming matrices of terrain elevations and adding a header reflecting the appropriate latitude and longitude coordinates corresponding to the INM study setup. The custom files were converted to the NED format for input into the INM framework.

In INM, adjustments are calculated based on the terrain features supplied in the data. A line-of-sight blockage adjustment is based on the theoretical barrier effect. It is computed and accounts for the difference in propagation path length between the direct path from source to receiver and the actual path from the source to receiver over the terrain feature(s). The calculation is a function of the Fresnel Number N_0, equal to the path length difference, normalized by ½ the wavelength of sound. Therefore, it is calculated independently for each 1/3 octave band and then logarithmically summed to obtain the full spectrum barrier effect[1]. A Lateral attenuation adjustment is also computed that accounts for attenuation due to ground, refraction-scattering, and engine installation effects. Only the larger of the two calculated adjustment terms—line-of-sight blockage or lateral attenuation—is used.

In addition, INM applies a study-wide adjustment for atmospheric absorption adjustment and acoustic impedance adjustment to the interpolated NPD. When the terrain feature is used, only the atmospheric absorption adjustment is applied study-wide. The acoustic impedance adjustment is applied to each receiver based on the receiver's altitude, temperature, and pressure to correct for the difference between receiver conditions and the reference day conditions that the NPD data are based on[1].

In the Nord2000 methodology, terrain is approximated by straight-line segments (with a limit of 1000 segments). Ground type and roughness must be defined for each segment. Only the two most efficient screens (terrain feature or man-made barrier) and the two most efficient edges of each screen are taken into account for the computation. The computation method is based on the concept of Fresnel-zones. Flat ground (with no terrain variation and one surface type) effects are calculated based on geometrical ray theory. Flat ground (with no terrain variation and multiple surface types) effects are calculated based on a modified Fresnel-zone method[7]. The effect of diffraction around terrain features is incorporated with diffraction coefficients, calculated as functions of geometrical parameters of the terrain shape and reflection coefficients of the terrain faces, as described in Section 1.

7.2 Terrain Comparisons

In order to approximate terrain using line segments, a diagnostic file was created to output geometrical data from an INM run. INM was run using NED terrain data which were then approximated in line segments for input into Nord2000.

Three different terrain features were compared including a downward slope, an upward slope, and a hill.

7.2.1 Terrain Comparison Input Parameters

The input parameters are the same for both INM and Nord2000 as both models allow for terrain input. The input parameters are described in Table 8. Parameters that have been changed from the baseline conditions are identified by italicized text.

Table 8. INM and Nord2000 Terrain Comparison Input Parameters

	Terrain – Downward Slope	Terrain – Upward Slope	Terrain - Hill
Temperature	15°C	15°C	15°C
Temperature Gradient	0	0	0
Relative Humidity	70%	70%	70%
Turbulence	INM: N/A Nord2000: 0	INM: N/A Nord2000: 0	INM: N/A Nord2000: 0
Terrain	*Downward Slope*	*Upward Slope*	*Hill*
Terrain Feature Height	*70 m*	*70 m*	*70 m*
Ground Type	Soft, 80 CGS Rayls	Soft, 80 CGS Rayls	Soft, 80 CGS Rayls

7.2.2 Terrain Comparison Results

Downward Slope Comparison

The results for the downward slope terrain comparison are shown in Table 9 and Figure 7. The LAMAX results are presented as the mean difference +/- the standard deviations of differences between INM and Nord2000 at the 11 receivers. The terrain and ground type are shown in Figure

8. The vertical dotted lines indicate ranges at which transitions in propagation conditions occur. Table 10 and Table 11 display the differences between each model's baseline and downward slope terrain case results.

With downward sloping terrain, Nord2000 computes LAMAX 4.8 +/- 2.2 dB greater than INM. The maximum difference between the two models for this comparison is 8.0 dB, at the farthest receiver. For this comparison, the 0-800 m results should match baseline conditions, but not 1200-3000 m. This is because the source height is defined above the terrain, which is elevated compared to the baseline. INM results show differences increasing with distance as expected. Nord2000 results mainly show the greatest effect just after the terrain slope transition (1200 m). In contrast to Nord2000, INM reports the greatest differences due to a change in elevation (1200-3000 m) rather than the effect of the sloping terrain. The engine installation effect and refraction-scattering components of the lateral attenuation adjustment may contribute to the difference in trend between INM and Nord2000.

Table 9. Downward Slope Terrain Results

Receptor Row Fourteen	Distance (m)	INM (dB): Downward Slope	Nord2000 (dB): Downward Slope	Difference (dB)
14_10	0	79.8	81.8	-2.0
14_9	300	75.7	77.9	-2.2
14_8	600	69.8	72.8	-3.0
14_7	900	65.2	68.8	-3.6
14_6	1200	62.0	65.3	-3.3
14_5	1500	58.6	63.0	-4.3
14_4	1800	55.8	61.1	-5.3
14_3	2100	53.2	59.5	-6.2
14_2	2400	51.0	58.0	-7.0
14_1	2700	49.0	56.6	-7.6
14_0	3000	47.2	55.2	-8.0

Table 10. Downward Slope Terrain Comparison, INM Difference from Baseline

Distance (m)	INM (dB): Baseline	INM (dB): Downward Slope	Difference (dB)
0	79.8	79.8	0.0
300	76.0	75.7	0.3
600	69.8	69.8	0.0
900	65.2	65.2	0.0
1200	61.5	62.0	-0.5
1500	58.0	58.6	-0.6
1800	55.0	55.8	-0.7
2100	52.4	53.2	-0.8
2400	50.2	51.0	-0.8
2700	48.1	49.0	-0.8
3000	46.3	47.2	-0.8

Table 11. Downward Slope Terrain Comparison, Nord2000 Difference from Baseline

Distance (m)	Nord2000 (dB): Baseline	Nord2000 (dB): Downward Slope	Difference (dB)
0	81.8	81.8	0.0
300	77.9	77.9	0.0
600	72.8	72.8	0.0
900	68.8	68.8	0.1
1200	65.6	65.3	0.3
1500	63.1	63.0	0.1
1800	61.1	61.1	0.0
2100	59.5	59.5	0.0
2400	58.0	58.0	0.0
2700	56.6	56.6	0.1
3000	55.3	55.2	0.1

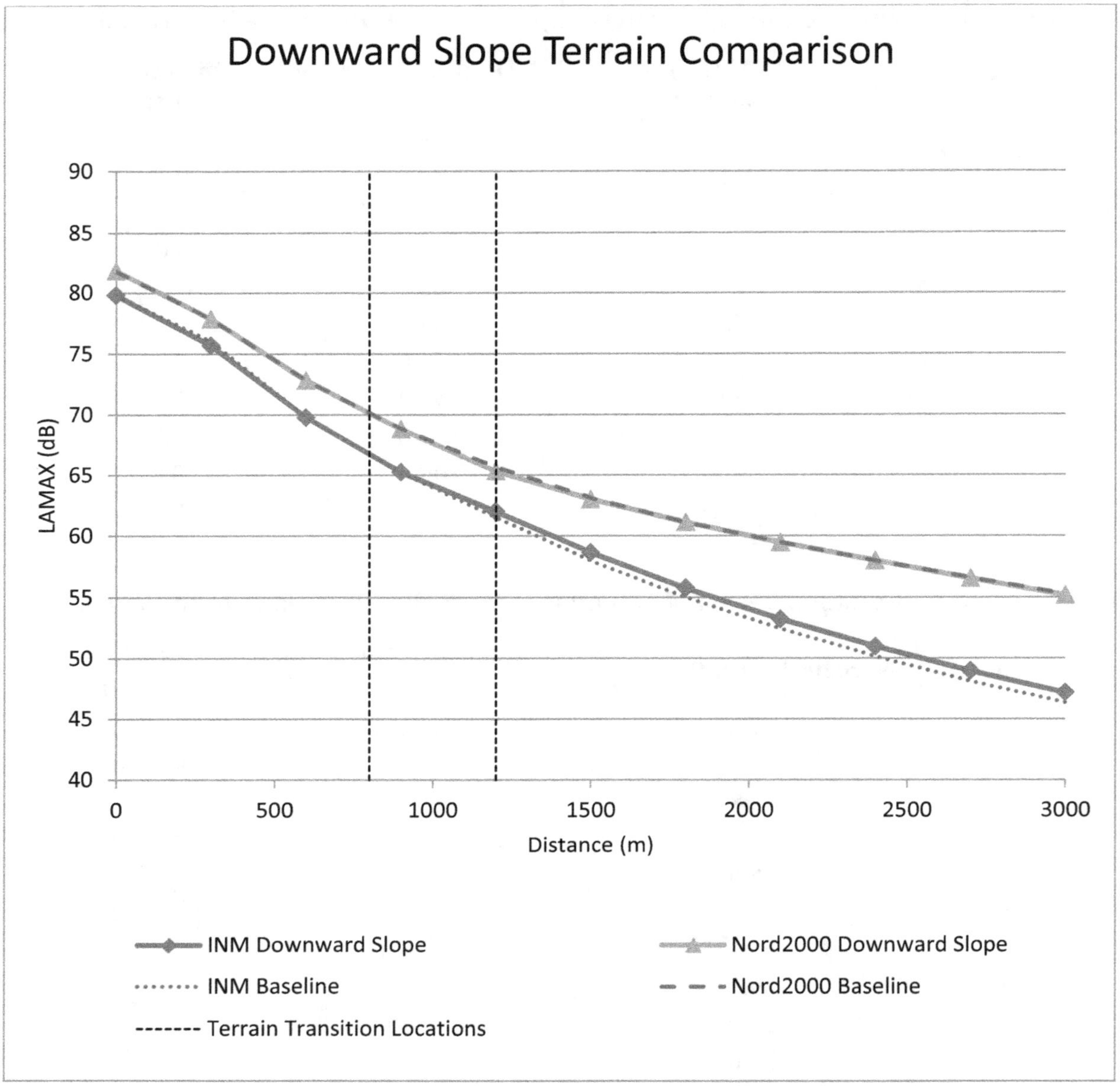

Figure 7. Downward Slope Terrain Results

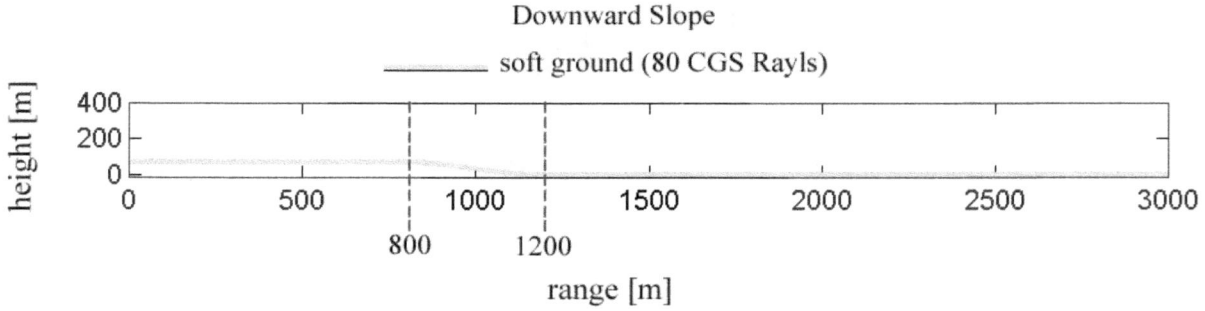

Figure 8. Downward Slope Terrain and Ground Type

Upward Slope Comparison

The results for the upward slope terrain comparison are shown in Table 12 and Figure 9. The LAMAX results are presented as the mean difference +/- the standard deviations of differences between INM and Nord2000 at the 11 receivers. The terrain and ground type are shown in Figure 10. The vertical dotted lines indicate ranges at which transitions in propagation conditions occur. Table 13 and Table 14 display the differences between the each model's baseline and upward slope terrain case results.

With upward sloping terrain, Nord2000 computes LAMAX 6.1 +/- 3.0 dB greater than INM. The maximum difference between the two models for this comparison is 10.1 dB, at the farthest receiver. The upward slope and baseline cases show similar results for Nord2000, within 0.4 dB for all receivers. However, INM results in the upward slope case are approximately 1 dB lower at the far receivers. Therefore, the largest difference between INM and Nord2000 at the farthest receiver is increased.

Table 12. Upward Slope Terrain Results

Receptor Row Fourteen	Distance (m)	INM (dB): Upward Slope	Nord2000 (dB): Upward Slope	Difference (dB)
14_10	0	79.8	81.8	-1.9
14_9	300	75.7	77.9	-2.1
14_8	600	69.8	72.8	-3.0
14_7	900	65.1	68.9	-3.8
14_6	1200	60.7	65.9	-5.3
14_5	1500	57.0	63.3	-6.3
14_4	1800	54.0	61.5	-7.5
14_3	2100	51.4	59.6	-8.2
14_2	2400	49.1	58.0	-9.0
14_1	2700	47.1	56.7	-9.6
14_0	3000	45.3	55.4	-10.1

Table 13. Upward Slope Terrain Comparison, INM Difference from Baseline

Distance (m)	INM (dB): Baseline	INM (dB): Upward Slope	Difference (dB)
0	79.8	79.8	0.0
300	76.0	75.7	0.3
600	69.8	69.8	0.0
900	65.2	65.1	0.1
1200	61.5	60.7	0.9
1500	58.0	57.0	1.0
1800	55.0	54.0	1.0
2100	52.4	51.4	1.1
2400	50.2	49.1	1.1
2700	48.1	47.1	1.0
3000	46.3	45.3	1.0

Table 14. Upward Slope Terrain Comparison, Nord2000 Difference from Baseline

Distance (m)	Nord2000 (dB): Baseline	Nord2000 (dB): Upward Slope	Difference (dB)
0	81.8	81.8	0.0
300	77.9	77.9	0.0
600	72.8	72.8	0.0
900	68.8	68.9	0.0
1200	65.6	65.9	-0.3
1500	63.1	63.3	-0.2
1800	61.1	61.5	-0.4
2100	59.5	59.6	-0.1
2400	58.0	58.0	0.0
2700	56.6	56.7	-0.1
3000	55.3	55.4	-0.1

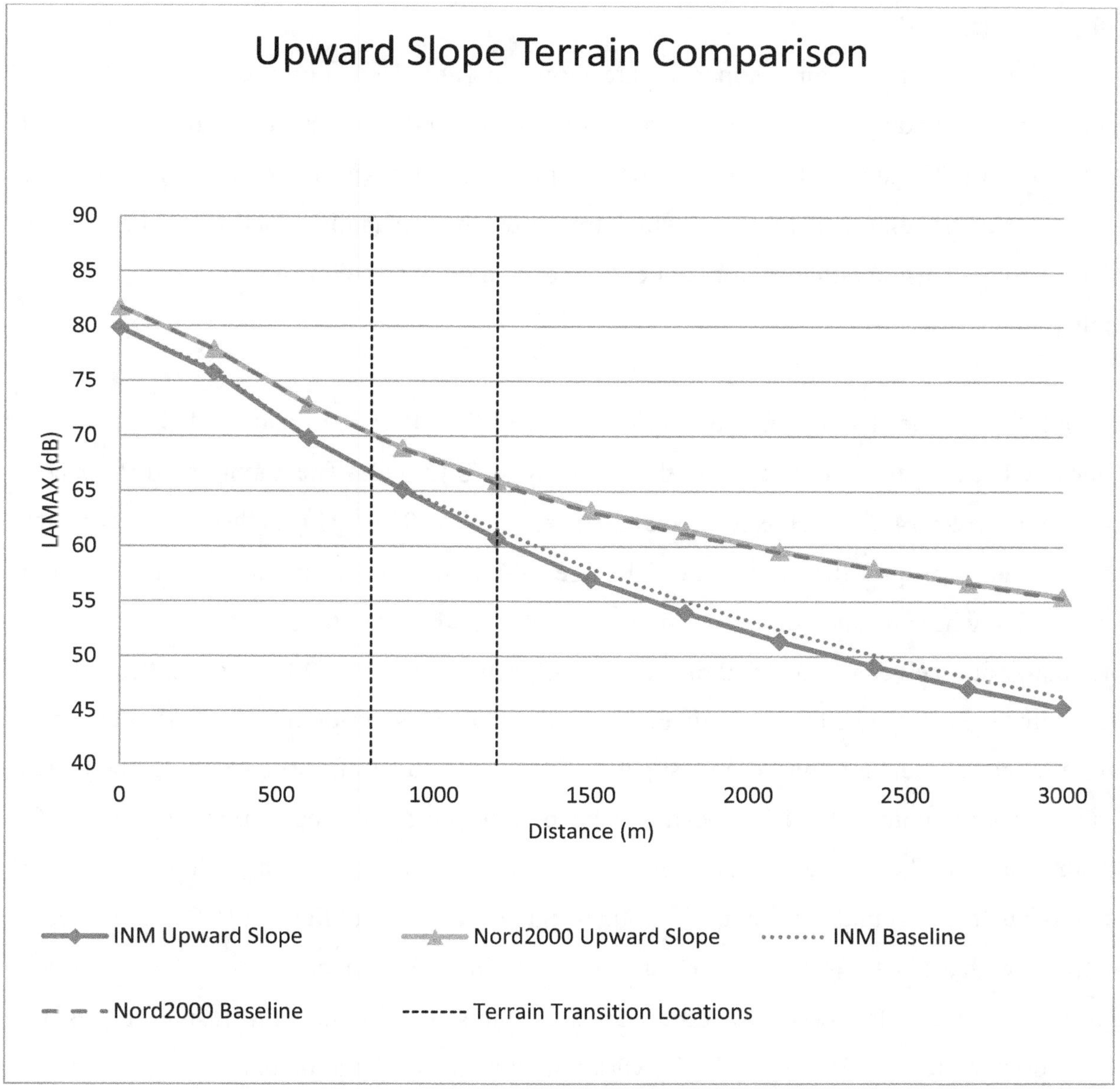

Upward Slope Terrain Comparison

Figure 9. Upward Slope Terrain Results

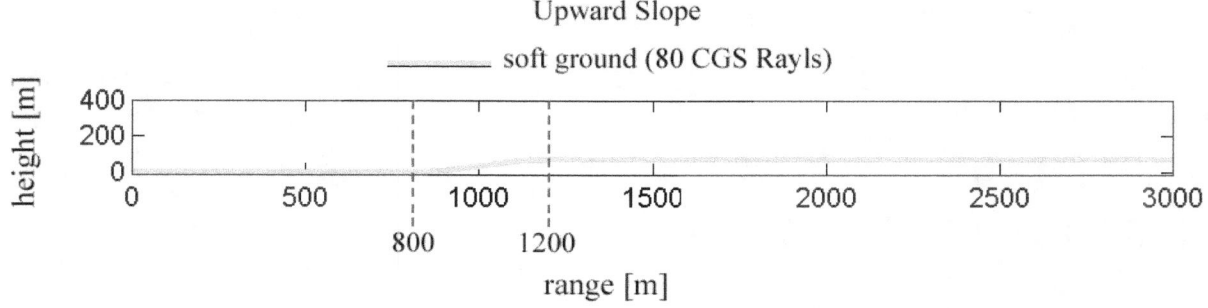

Figure 10. Upward Slope Terrain and Ground Type

Hill Comparison

The results for the hill terrain comparison are shown in Table 15 and Figure 11. The LAMAX results are presented as the mean difference +/- the standard deviations of differences between INM and Nord2000 at the 11 receivers. The terrain and ground type are shown in Figure 12. The vertical dotted lines indicate ranges at which transitions in propagation conditions occur. Table 16 and Table 17 display the differences between each model's baseline and hill terrain case results.

With hill terrain, Nord2000 computes LAMAX 6.4 +/- 4.0 dB greater than INM. The maximum difference between the two models for this comparison is 15.6 dB. The maximum difference occurs at receiver 14_5 which is located at approximately 1500 m. This is the first receiver to report sound levels past the peak of the hill. Here a dip in level of more than 8 dB is seen in the INM results, where no such drop is seen in Nord2000 results. The increase in sound level, particularly at receiver 14_5 (1500 m), was not expected in the Nord2000 results. Possible sources for the unexpected trend could be due to the coarse resolution of receivers and the small size of the terrain feature. Further investigation is needed to determine the source of the increase. It is important to note that the hill location is beyond the validated propagation distance of 400 m for non-flat terrain. Other research currently being done on a hybrid propagation model (HPM) also models these ground conditions[10,11]. The results for these conditions in HPM, presented in Figure 13 with a 1310 ft (400 m) source altitude, show a slight increase in noise level before a significant decrease, followed by another increase in noise level,. The HPM results suggest that Nord2000 may be experiencing a similar variation in noise level and that the resolution of the receivers does not accurately capture the full effect of the hill terrain feature.

Table 15. Hill Terrain Results

Receptor Row Fourteen	Distance (m)	INM (dB): Hill	Nord2000 (dB): Hill	Difference (dB)
14_10	0	79.8	81.8	-1.9
14_9	300	75.7	77.9	-2.1
14_8	600	69.8	72.8	-3.0
14_7	900	65.1	68.9	-3.8
14_6	1200	60.7	65.9	-5.3
14_5	1500	49.4	65.0	-15.6
14_4	1800	55.0	61.4	-6.3
14_3	2100	52.4	59.7	-7.2
14_2	2400	50.2	58.1	-8.0
14_1	2700	48.1	56.7	-8.6
14_0	3000	46.3	55.3	-9.0

Table 16. Hill Terrain Comparison, INM Difference from Baseline

Distance (m)	INM (dB): Baseline	INM (dB): Hill	Difference (dB)
0	79.8	79.8	0.0
300	76.0	75.7	0.3
600	69.8	69.8	0.0
900	65.2	65.1	0.1
1200	61.5	60.7	0.9
1500	58.0	49.4	8.6
1800	55.0	55.0	0.0
2100	52.4	52.4	0.0
2400	50.2	50.2	0.0
2700	48.1	48.1	0.0
3000	46.3	46.3	0.0

Table 17. Hill Terrain Comparison, Nord2000 Difference from Baseline

Distance (m)	Nord2000 (dB): Baseline	Nord2000 (dB): Hill	Difference (dB)
0	81.8	81.8	0.0
300	77.9	77.9	0.0
600	72.8	72.8	0.0
900	68.8	68.9	0.0
1200	65.6	65.9	-0.3
1500	63.1	65.0	-1.9
1800	61.1	61.4	-0.3
2100	59.5	59.7	-0.2
2400	58.0	58.1	-0.1
2700	56.6	56.7	-0.1
3000	55.3	55.3	0.0

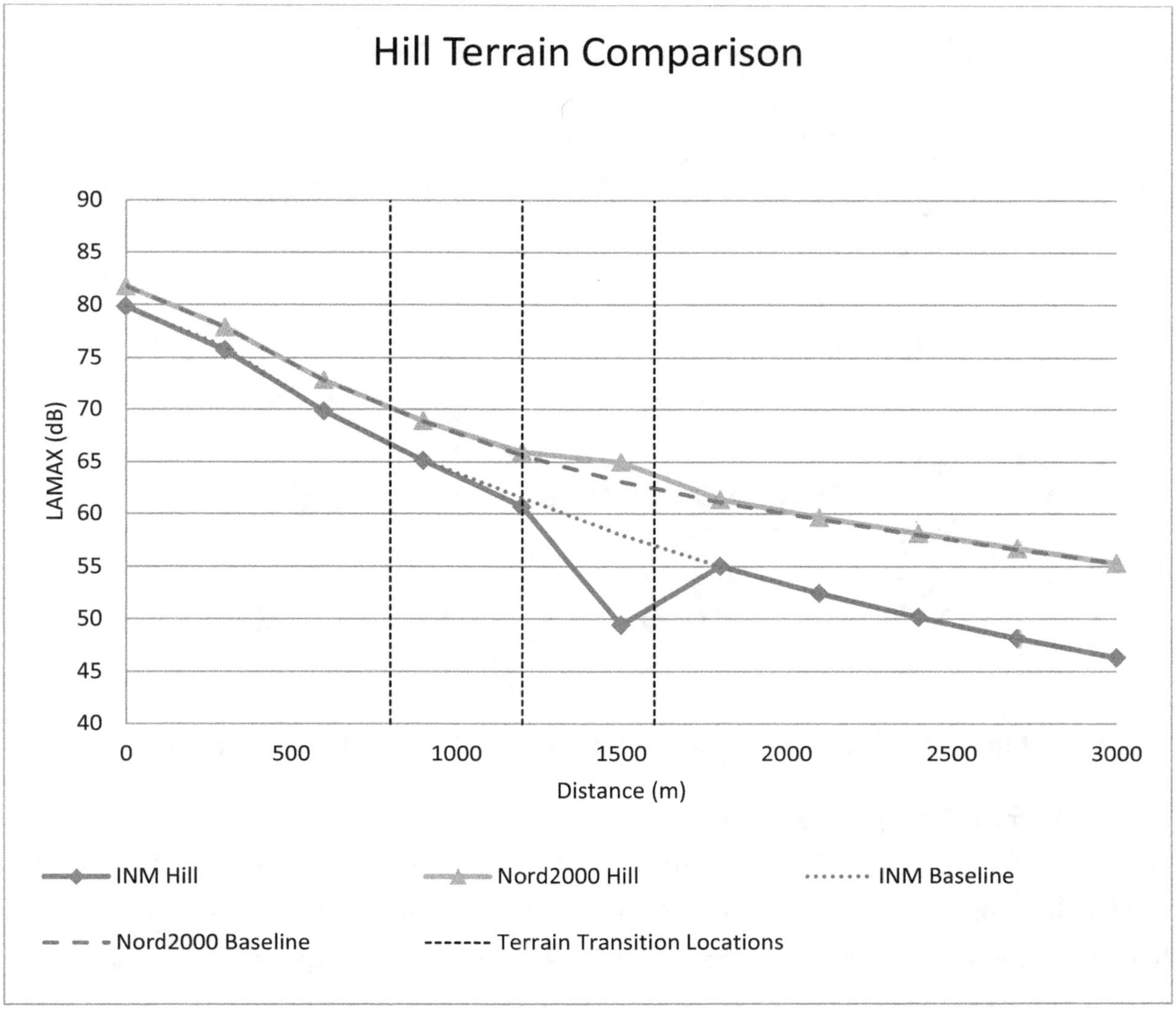

Hill Terrain Comparison

[Graph showing LAMAX (dB) vs Distance (m) from 0 to 3000m, comparing INM Hill, Nord2000 Hill, INM Baseline, Nord2000 Baseline, with Terrain Transition Locations at 800, 1200, and 1600m]

Figure 11. Hill Terrain Results

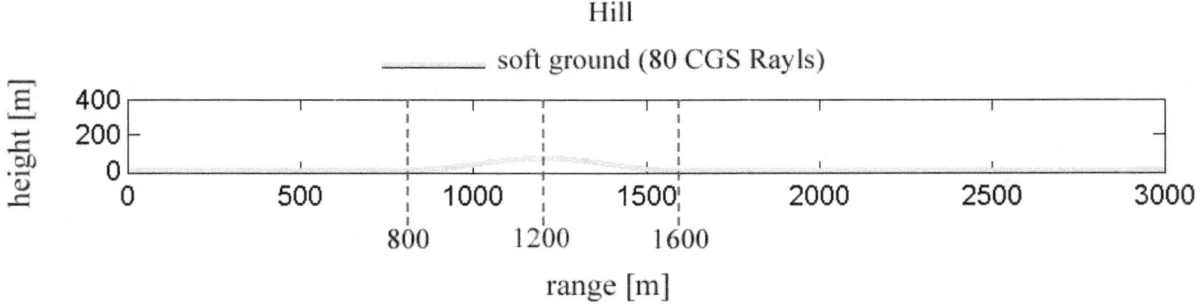

Figure 12. Hill Terrain and Ground Type

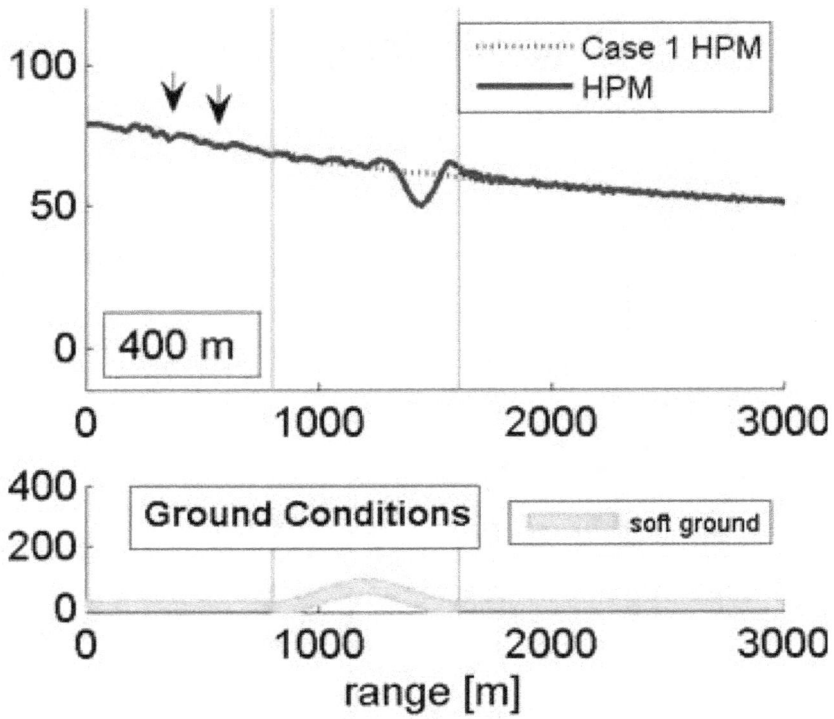

Figure 13. Hill Terrain Results from Hybrid Propagation Research[10]

7.2.3 Terrain Comparison Summary

Figure 14 shows all 3 terrain feature comparisons against the baseline of each model in order to highlight the magnitude of the effect of each terrain feature on the LAMAX results. The hill case shows the greatest variation in trend between INM and Nord2000 as expected, but as mentioned above the increase in sound level at the hill terrain feature was not expected in the Nord2000 results. Further investigation is needed to determine the source of the increase.

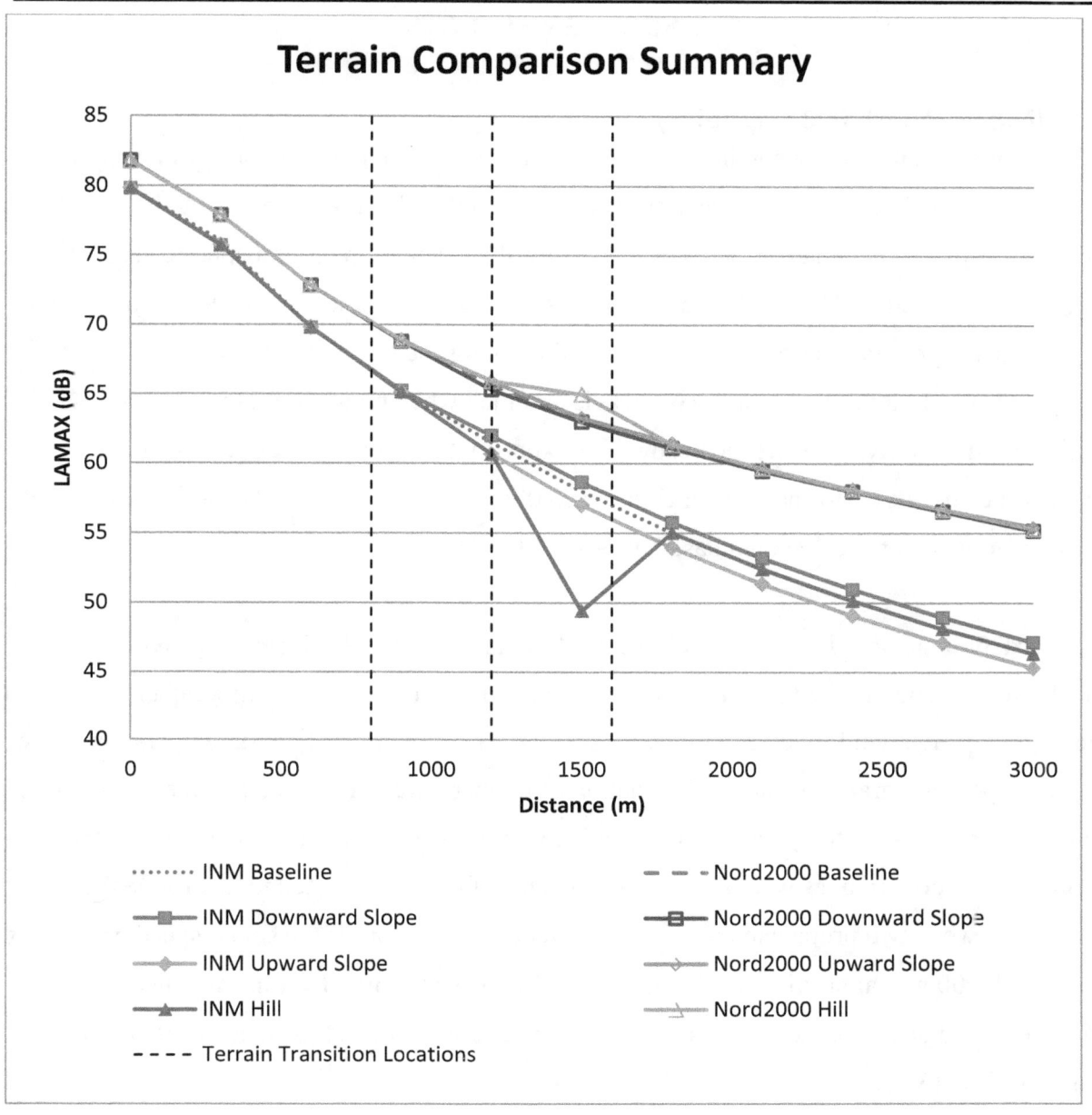

Figure 14. Terrain Comparison Summary Graph

8 WEATHER EFFECTS

8.1 INM vs. Nord2000 Capability

INM uses average annual day conditions for temperature, relative humidity, and atmospheric pressure for the study airport. These conditions are user defined and are intended to be representative of a typical day at the study airport for the calculation of atmospheric absorption and acoustic impedance. The INM default airport standard day conditions are 59 °F temperature and 29.92 in-Hg pressure, based on sea-level conditions for the International Standard Atmosphere (ISA). Relative humidity of 70% is assumed when no user input is provided. The user also has the ability to define the headwind in knots, which indirectly effects noise by impacting the aircraft performance, which in turn affects the thrust levels. The atmospheric absorption adjustment is calculated based on SAE-ARP-866A.

In the Nord2000 methodology, meteorological effects are calculated using an approximate vertical effective sound speed profile. The user has control over the following weather parameters: average wind speed at a given height, standard deviation of wind, temperature at the ground, temperature gradient, standard deviation of temperature, turbulence parameter for wind, turbulence parameter for temperature, and relative humidity. These parameters allow for the computation of refraction, as well as some coefficients of coherence, used to determine the coherence between two propagated rays. In the event of refraction (if the sound speed gradient is not 0), Nord2000 will approximate a logarithmic sound speed profile by a linear sound speed profile and propagate with curved rays. An atmospheric absorption adjustment is calculated based on ISO 9613-1[3].

Weather classes can be established in the Nord2000 methodology based on wind speed and atmospheric stability. These can be used to determine the effective sound speed profile[2]. Twenty-five classes have been established, but not implemented in the version of the Nord2000 code obtained in the software exchange. Statistical weights can be determined for how often each weather class occurs over a period of time at a given location. More research should be done on this topic, but preliminary findings suggest that implementing weather classes in AEDT/INM would be beneficial.

8.2 Atmospheric Absorption Comparisons

Atmospheric absorption comparisons were done outside of the full Nord2000 and INM implementations using only the atmospheric absorption standards. An in-house tool was developed to run each atmospheric absorption standard with INM spectral class data input. The current atmospheric absorption standard used in INM is SAE-ARP-866A. The current atmospheric absorption standard used in the Nord2000 methodology is ISO 9613-1. SAE-ARP-5534 is a pending replacement for SAE-ARP-866A in INM, so it was also compared.

8.2.1 Atmospheric Absorption Conditions

Comparisons were done between the three standards under the following conditions:

- High Humidity (90%)
    - High Temperature (90°F)
    - Standard Temperature (59°F)
    - Low Temperature (40°F)
- Standard Humidity (70%)
    - High Temperature (90°F)
    - Standard Temperature (59°F)
    - Low Temperature (40°F)
- Intermediate Humidity (45%)
    - High Temperature (90°F)
    - Standard Temperature (59°F)
    - Low Temperature (40°F)
- Low Humidity (20%)
    - High Temperature (90°F)
    - Standard Temperature (59°F)
    - Low Temperature (40°F)

8.2.2 Atmospheric Absorption Comparison Results

The results reported in the following graphs are averaged over all INM spectral classes for each slant distance. The standard deviation is shown in the error bars on the graph. The data are presented in tabular form in Appendix C.

High Humidity Comparison

The results for the high humidity case are shown in Figure 15. Note that all three approaches show close agreement.

Standard Humidity Comparison

The results for the standard humidity case are shown in Figure 16. Note that all approaches show close agreement.

Intermediate Humidity Comparison

The results for the intermediate humidity case are shown in Figure 17. Note that all three approaches show close agreement with the exception of the SAE-ARP-866A 90°F case.

Low Humidity Comparison

The results for the Low humidity case are shown in Figure 18. Note that all three approaches show close agreement with the exception of the SAE-ARP-866A 40°F case.

Overall, SAE-ARP 5534, the pending replacement for SAE-ARP-866A in INM, is in much better agreement with Nord2000's ISO 9613-1 atmospheric absorption standard than the current INM atmospheric absorption standard. Note that all other comparisons for this research were done with the current INM atmospheric absorption, SAE-ARP-866A. ISO 9613-1 predictions for the standard humidity and temperature are typically higher than those of SAE-ARP-866A, just as Nord2000 typically predicts higher levels than INM. The difference in atmospheric absorption standards does not account for the entire difference between INM and Nord2000. However, it is one contributor.

Weather Effects

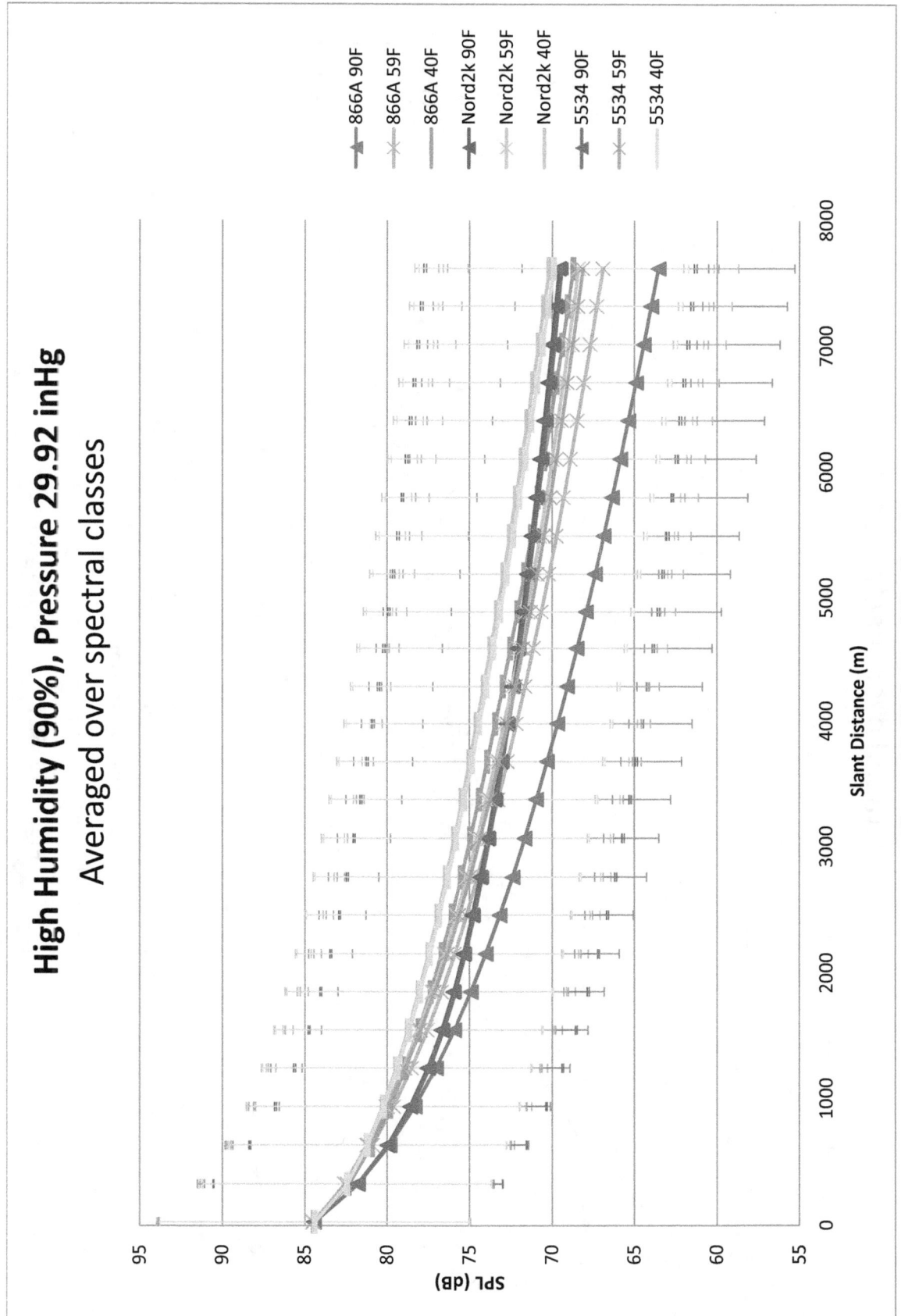

Figure 15. High Humidity Results

Weather Effects

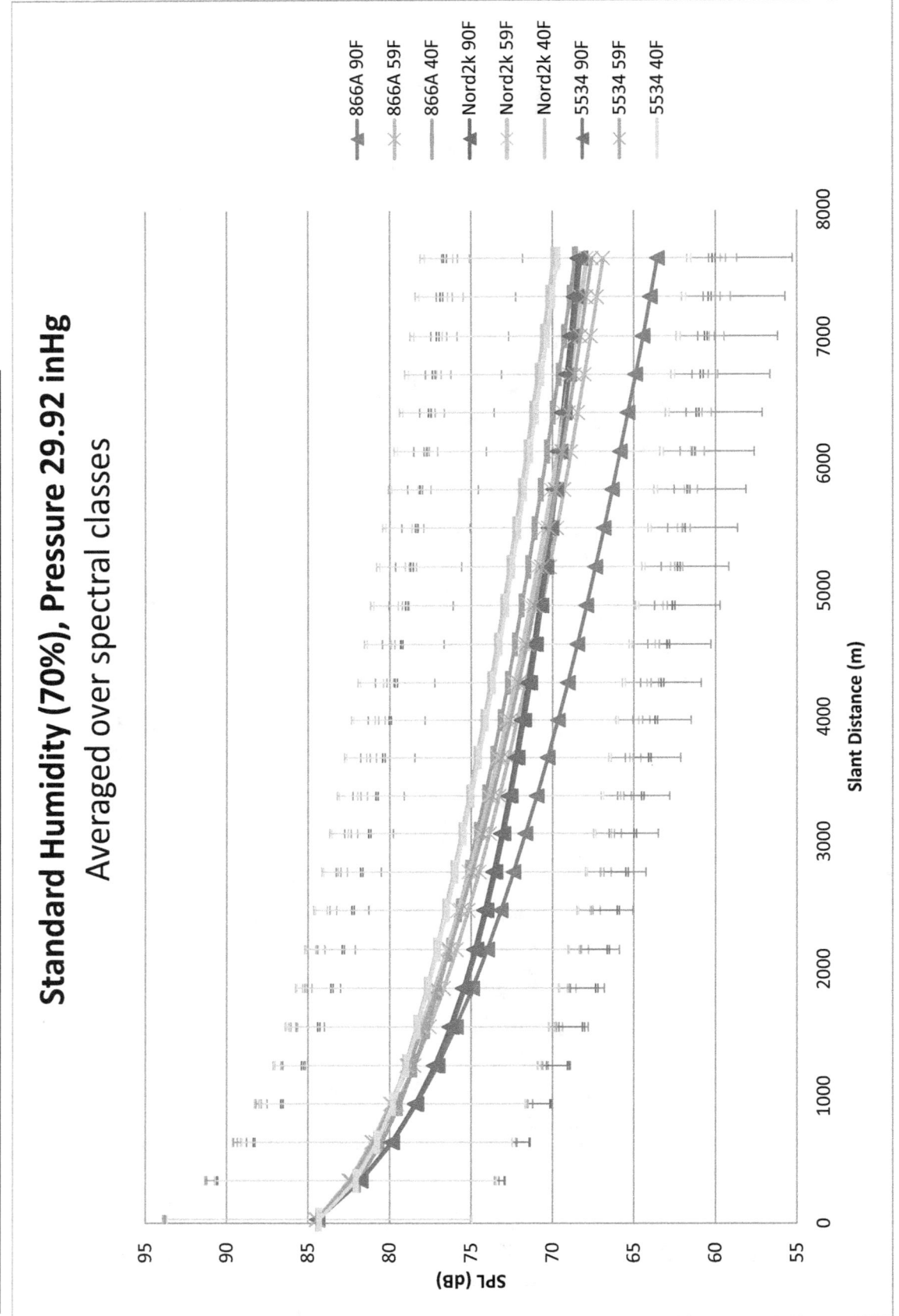

Figure 16. Standard Humidity Results

Weather Effects

The Analysis of Modeling Aircraft Noise with the Nord2000 Noise Model

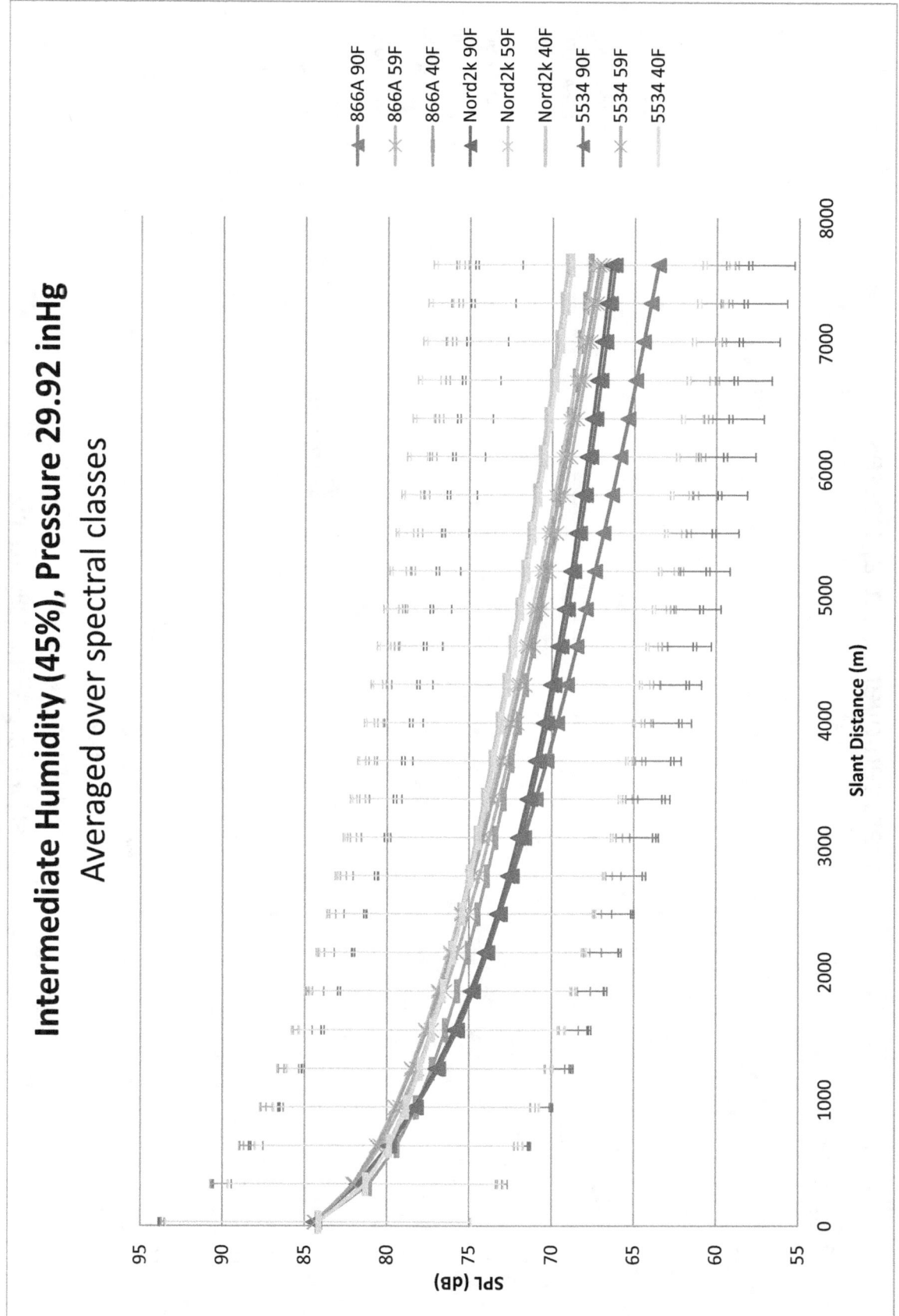

Figure 17. Intermediate Humidity Results

Weather Effects

The Analysis of Modeling Aircraft Noise with the Nord2000 Noise Model

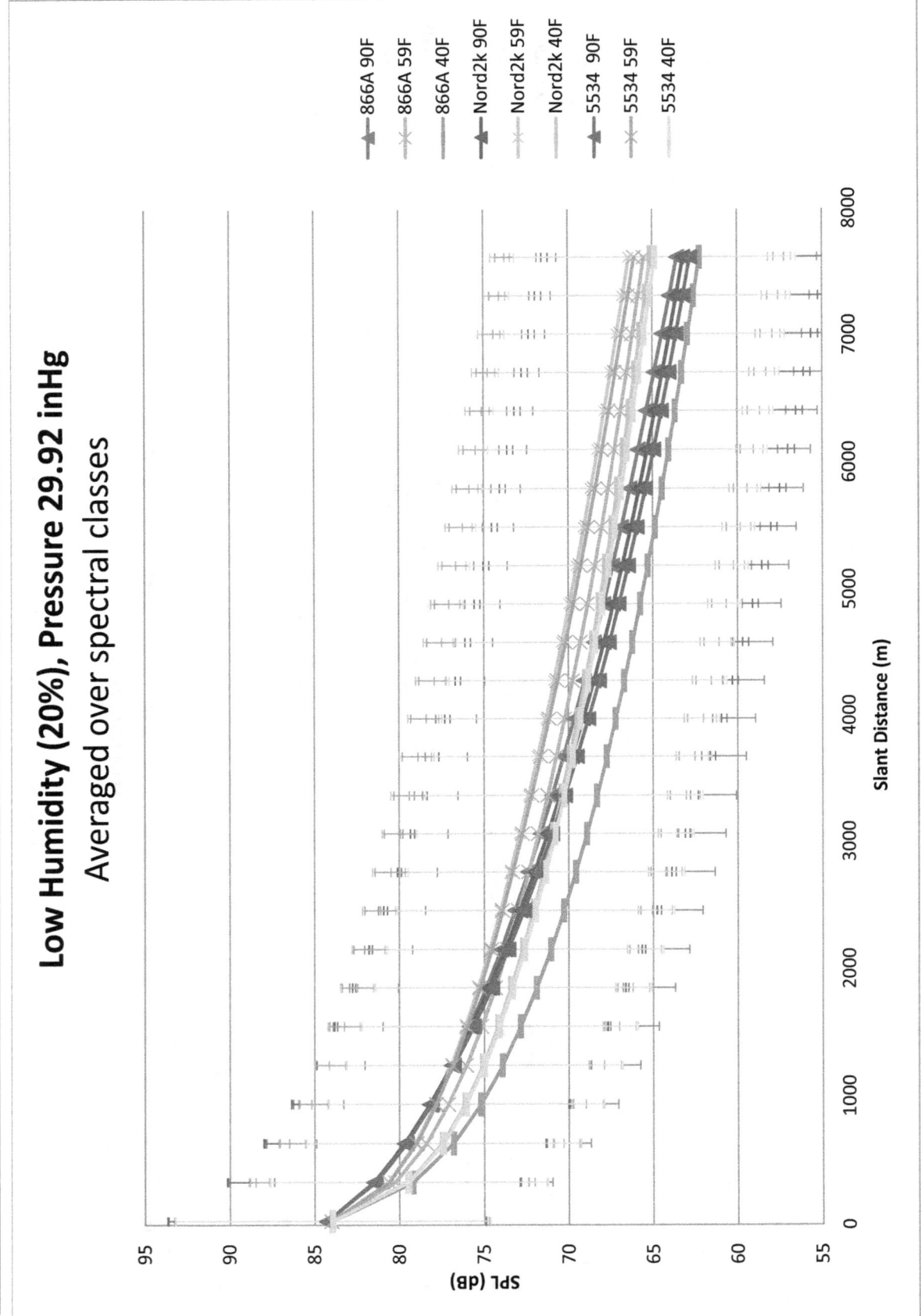

Figure 18. Low Humidity Results

8.3 Atmospheric Profile Comparisons

Atmospheric profile scenarios were compared for the following six conditions:

- High Humidity, High Temperature
- High Humidity, Low Temperature
- Low Humidity, High Temperature
- Low Humidity, Low Temperature
- Positive Temperature Gradient
- Negative Temperature Gradient

8.3.1 Atmospheric Profile Input Parameters

Four out of the six comparisons were done with like input parameters as both INM and Nord2000 allow for humidity and temperature. INM does not allow for a temperature gradient input so the Nord2000 output was compared to the INM baseline output for those scenarios. The input parameters are shown in Table 18 and Table 19. Parameters that were changed from the baseline conditions are identified by italicized text.

Table 18. INM and Nord2000 Atmospheric Profile Comparison Input Parameters

	High Humid/Temp	High Humid Low Temp	Low Humid High Temp	Low Humid/Temp
Temperature	*32.2°C*	*4.4°C*	*32.2°C*	*4.4°C*
Temperature Gradient	0	0	0	0
Relative Humidity	*90%*	*90%*	*20%*	*20%*
Turbulence	INM: N/A Nord2000: 0	INM: N/A Nord2000: 0	INM: N/A Nord2000: 0	INM: N/A Nord2000: 0
Terrain	Flat	Flat	Flat	Flat
Ground Type	Soft, 80 CGS Rayls	Soft, 80 CGS Rayls	Soft, 80 CGS Rayls	Soft, 80 CGS Rayls

Table 19. Nord2000 Atmospheric Profile Comparison Input Parameters, INM Baseline

	Positive Temp Gradient	Negative Temp Gradient
Temperature	15°C	15°C
Temperature Gradient	*0.2032 °C/m*	*-0.2032 °C/m*
Relative Humidity	70%	70%
Turbulence	INM: N/A Nord2000: 0	INM: N/A Nord2000: 0
Terrain	Flat	Flat
Ground Type	Soft, 80 CGS Rayls	Soft, 80 CGS Rayls

8.3.2 Homogeneous Atmosphere Comparison Results
High Humidity, High Temperature Comparison

The results for the high humidity, high temperature case are shown in Table 20 and Figure 19. The LAMAX results are presented as the mean difference +/- the standard deviations of differences between INM and Nord2000 at the 11 receivers. The terrain and ground type are shown in Figure 20. Table 21 and Table 22 display the differences between each model's baseline and high humidity, high temperature case results.

Nord2000 computes LAMAX 5.4 +/- 2.9 dB greater than INM. The maximum difference between the two models for this comparison is 9.4 dB, at the farthest receiver. The differences between each model's baseline and high humidity, high temperature case (Table 21 and Table 22) show a similar trend in INM and Nord2000, with a maximum difference of 0.5 dB.

Table 20. High Humidity, High Temperature Results

Receptor Row Fourteen	Distance (m)	INM (dB): High Humidity, High Temp	Nord2000 (dB): High Humidity, High Temp	Difference (dB)
14_10	0	79.1	80.8	-1.7
14_9	300	75.1	76.7	-1.6
14_8	600	68.6	71.3	-2.7
14_7	900	63.8	67.2	-3.4
14_6	1200	59.9	63.9	-4.0
14_5	1500	56.2	61.5	-5.3
14_4	1800	53.1	59.5	-6.4
14_3	2100	50.3	57.8	-7.5
14_2	2400	47.9	56.2	-8.3
14_1	2700	45.7	54.7	-8.9
14_0	3000	43.8	53.2	-9.4

Table 21. High Humidity, High Temperature Comparison, INM Difference from Baseline

Distance (m)	INM (dB): Baseline	INM (dB): High Humidity, High Temp	Difference (dB)
0	79.8	79.1	0.7
300	76.0	75.1	0.9
600	69.8	68.6	1.2
900	65.2	63.8	1.4
1200	61.5	59.9	1.6
1500	58.0	56.2	1.8
1800	55.0	53.1	2.0
2100	52.4	50.3	2.1
2400	50.2	47.9	2.3
2700	48.1	45.7	2.4
3000	46.3	43.8	2.5

Table 22. High Humidity, High Temperature Comparison, Nord2000 Difference from Baseline

Distance (m)	Nord2000 (dB): Baseline	Nord2000 (dB): High Humidity, High Temp	Difference (dB)
0	81.8	80.8	0.9
300	77.9	76.7	1.1
600	72.8	71.3	1.5
900	68.8	67.2	1.7
1200	65.6	63.9	1.7
1500	63.1	61.5	1.6
1800	61.1	59.5	1.6
2100	59.5	57.8	1.7
2400	58.0	56.2	1.8
2700	56.6	54.7	2.0
3000	55.3	53.2	2.1

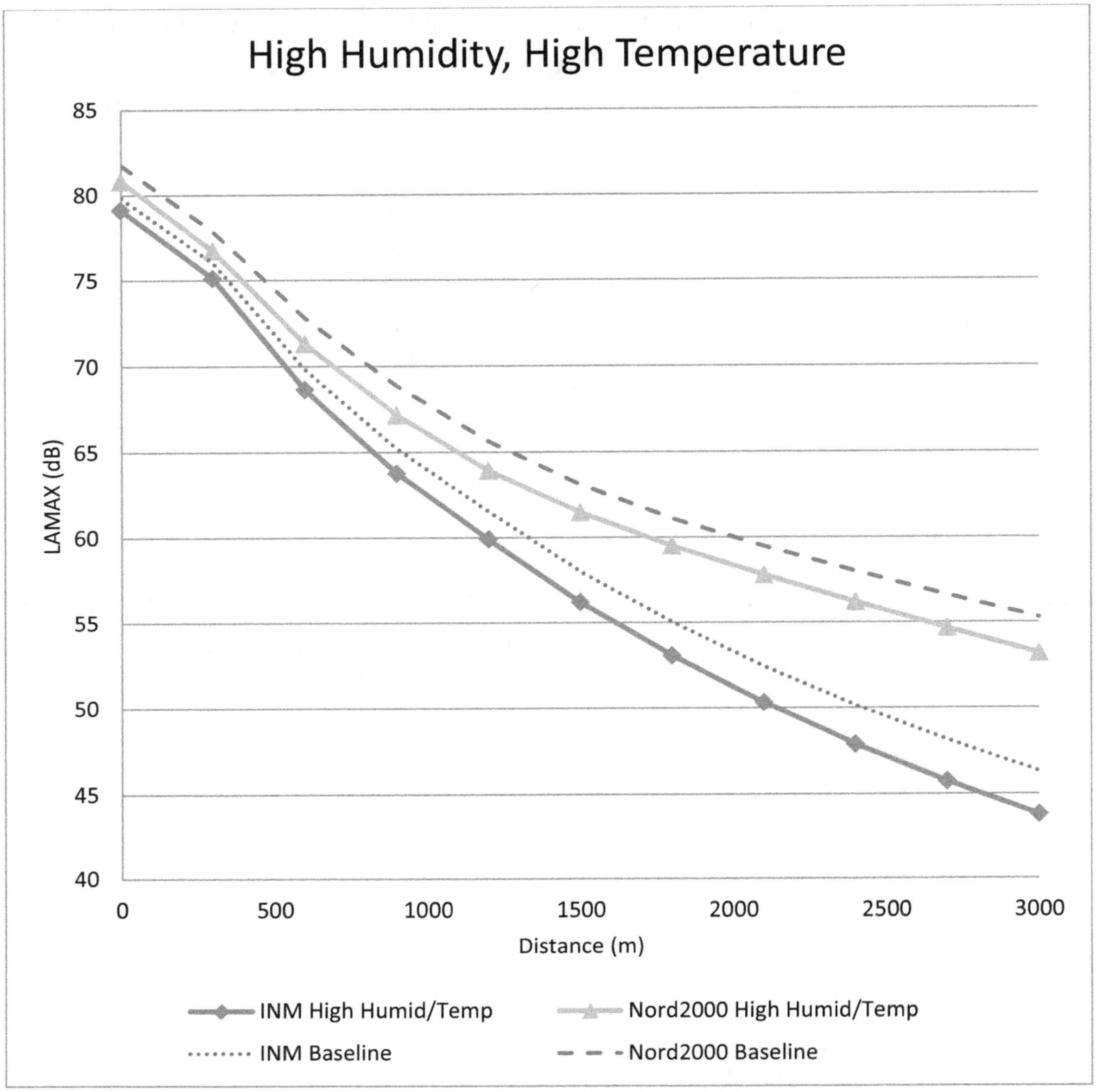

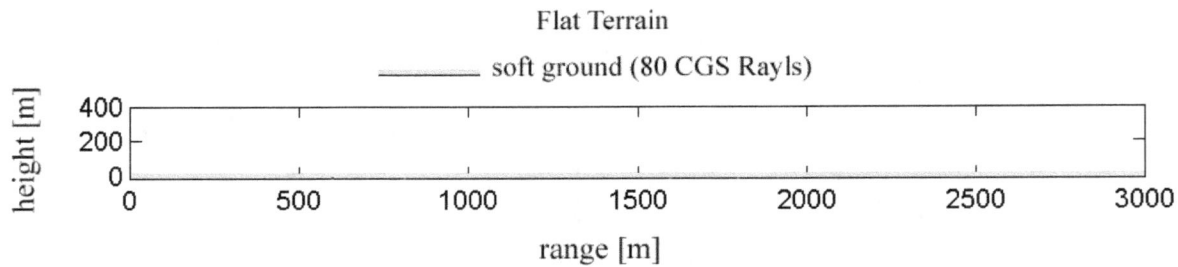

Figure 19. High Humidity, High Temperature Results

Figure 20. High Humidity, High Temperature Comparison Terrain and Ground Type

High Humidity, Low Temperature Comparison

The results for the high humidity, low temperature comparison are shown in Table 23 and Figure 21. The LAMAX results are presented as the mean difference +/- the standard deviations of differences between INM and Nord2000 at the 11 receivers. The terrain and ground type are shown in Figure 22. Table 24 and Table 25 display the differences between each model's baseline and high humidity, low temperature case results.

Nord2000 computes LAMAX 5.4 +/- 3.0 dB greater than INM. The maximum difference between the two models for this comparison is 9.8 dB, at the farthest receiver. The differences between each model's baseline and high humidity, low temperature case (Table 24 and Table 25) show a similar trend in INM and Nord2000, with a maximum difference of 0.7 dB.

Table 23. High Humidity, Low Temperature Results

Receptor Row Fourteen	Distance (m)	INM (dB): High Humidity, Low Temp	Nord2000 (dB): High Humidity, Low Temp	Difference (dB)
14_10	0	79.9	81.6	-1.7
14_9	300	76.2	77.7	-1.5
14_8	600	70.1	72.9	-2.7
14_7	900	65.7	69.1	-3.4
14_6	1200	62.1	66.2	-4.0
14_5	1500	58.7	63.9	-5.1
14_4	1800	55.9	62.2	-6.3
14_3	2100	53.4	60.8	-7.4
14_2	2400	51.2	59.6	-8.3
14_1	2700	49.3	58.4	-9.1
14_0	3000	47.6	57.3	-9.8

Table 24. High Humidity, Low Temperature Comparison, INM Difference from Baseline

Distance (m)	INM (dB): Baseline	INM (dB): High Humidity, Low Temp	Difference (dB)
0	79.8	79.9	-0.1
300	76.0	76.2	-0.2
600	69.8	70.1	-0.3
900	65.2	65.7	-0.5
1200	61.5	62.1	-0.6
1500	58.0	58.7	-0.7
1800	55.0	55.9	-0.9
2100	52.4	53.4	-1.0
2400	50.2	51.2	-1.1
2700	48.1	49.3	-1.2
3000	46.3	47.6	-1.3

Table 25. High Humidity, Low Temperature Comparison, Nord2000 Difference from Baseline

Distance (m)	Nord2000 (dB): Baseline	Nord2000 (dB): High Humidity, Low Temp	Difference (dB)
0	81.8	81.6	0.1
300	77.9	77.7	0.2
600	72.8	72.9	0.0
900	68.8	69.1	-0.2
1200	65.6	66.2	-0.5
1500	63.1	63.9	-0.8
1800	61.1	62.2	-1.1
2100	59.5	60.8	-1.3
2400	58.0	59.6	-1.6
2700	56.6	58.4	-1.8
3000	55.3	57.3	-2.1

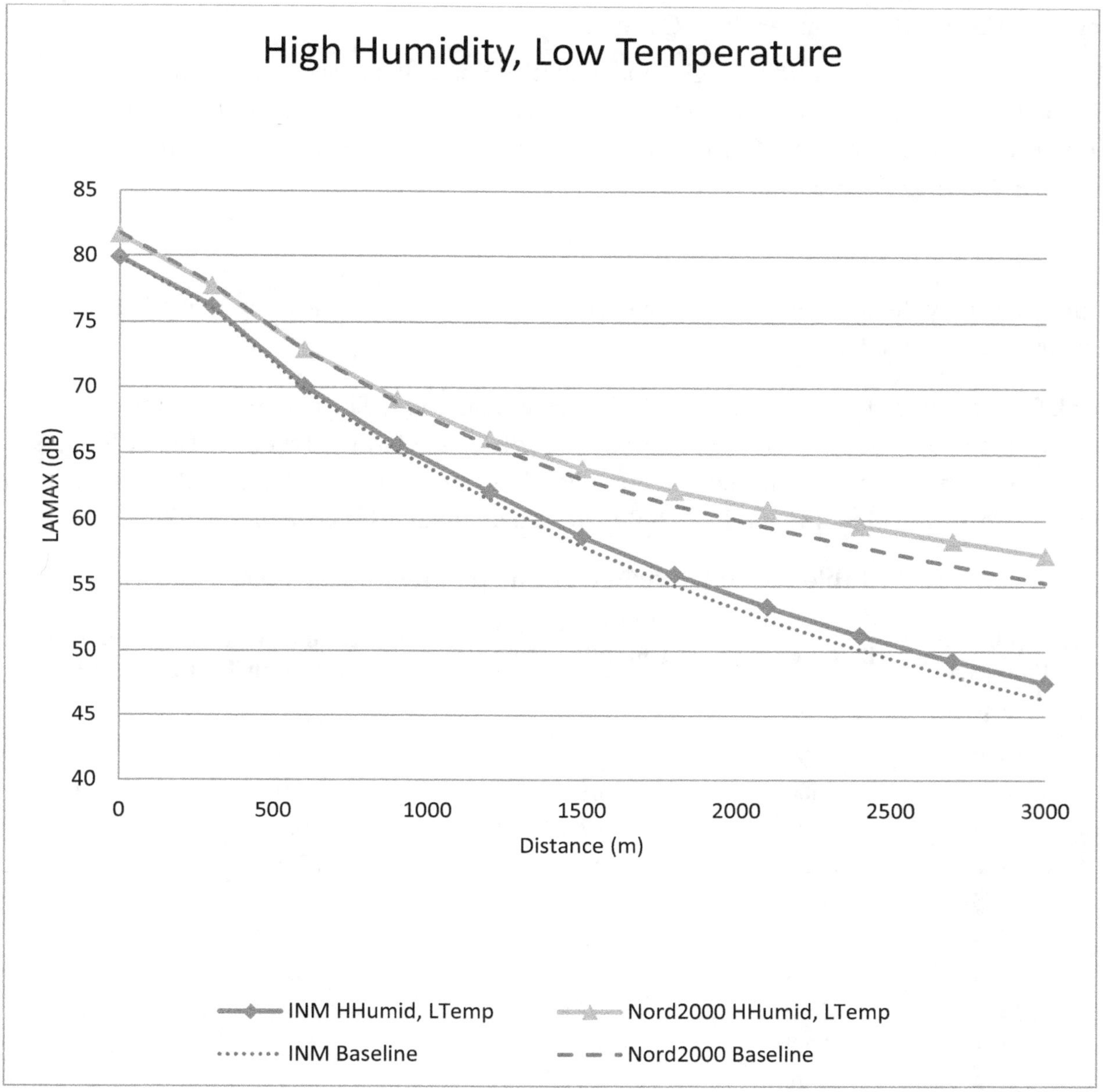

Weather Effects

The Analysis of Modeling Aircraft Noise with the Nord2000 Noise Model

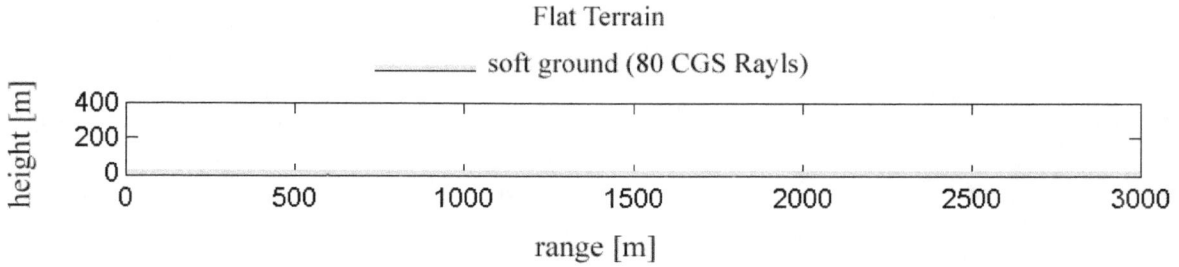

Figure 21. High Humidity, Low Temperature Results

Figure 22. High Humidity, Low Temperature Comparison Terrain and Ground Type

Low Humidity, High Temperature Comparison

The results for the low humidity, high temperature comparison are shown in Table 26 and Figure 23. The LAMAX results are presented as the mean difference +/- the standard deviations of differences between INM and Nord2000 at the 11 receivers. The terrain and ground type are shown in Figure 24. Table 27 and

Table 28 display the differences between each model's baseline and low humidity, high temperature case results.

Nord2000 computes LAMAX 4.5 +/- 2.1 dB greater than INM. The maximum difference between the two models for this comparison is 7.4 dB, at the farthest receiver. The differences between each model's baseline and low humidity, low temperature case (Table 27 and Table 28) show a similar trend in INM and Nord2000 with a maximum difference of 1.6 dB.

Table 26. Low Humidity, High Temperature Results

Receptor Row Fourteen	Distance (m)	INM (dB): Low Humidity, High Temp	Nord2000 (dB): Low Humidity, High Temp	Difference (dB)
14_10	0	78.8	80.5	-1.7
14_9	300	74.9	76.5	-1.6
14_8	600	68.5	71.0	-2.5
14_7	900	63.7	66.7	-3.1
14_6	1200	59.8	63.2	-3.4
14_5	1500	56.1	60.5	-4.3
14_4	1800	53.0	58.3	-5.2
14_3	2100	50.3	56.3	-6.1
14_2	2400	47.9	54.6	-6.7
14_1	2700	45.7	52.8	-7.1
14_0	3000	43.8	51.2	-7.4

Table 27. Low Humidity, High Temperature Comparison, INM Difference from Baseline

Distance (m)	INM (dB): Baseline	INM (dB): Low Humidity, High Temp	Difference (dB)
0	79.8	78.8	1.0
300	76.0	74.9	1.2
600	69.8	68.5	1.4
900	65.2	63.7	1.5
1200	61.5	59.8	1.7
1500	58.0	56.1	1.9
1800	55.0	53.0	2.0
2100	52.4	50.3	2.1
2400	50.2	47.9	2.3
2700	48.1	45.7	2.4
3000	46.3	43.8	2.5

Table 28. Low Humidity, High Temperature Comparison, Nord2000 Difference from Baseline

Distance (m)	Nord2000 (dB): Baseline	Nord2000 (dB): Low Humidity, High Temp	Difference (dB)
0	81.8	80.5	1.2
300	77.9	76.5	1.4
600	72.8	71.0	1.8
900	68.8	66.7	2.1
1200	65.6	63.2	2.4
1500	63.1	60.5	2.6
1800	61.1	58.3	2.8
2100	59.5	56.3	3.1
2400	58.0	54.6	3.4
2700	56.6	52.8	3.8
3000	55.3	51.2	4.1

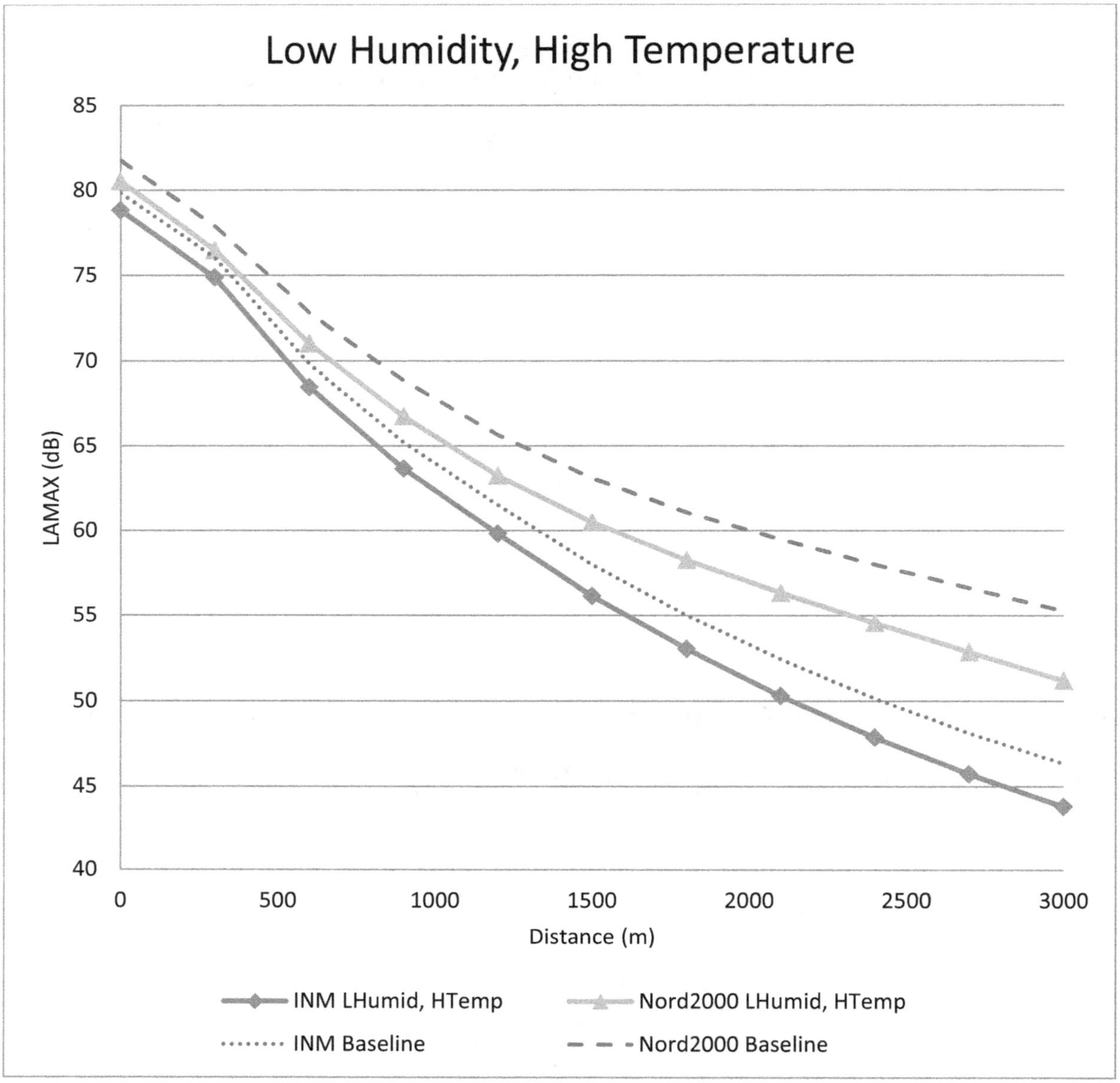

Figure 23. Low Humidity, High Temperature Results

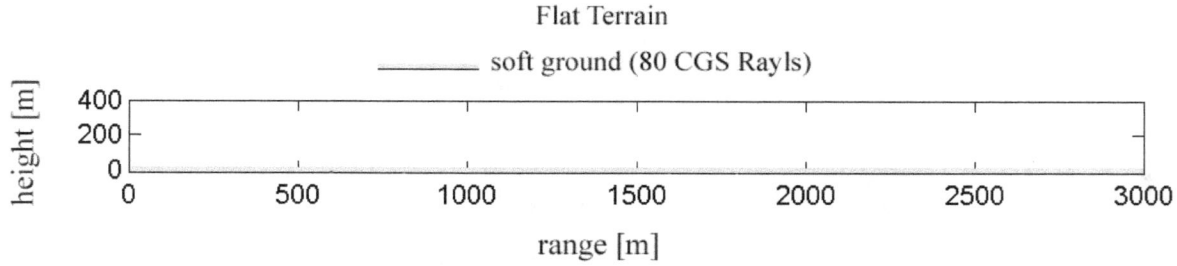

Figure 24. Low Humidity, High Temperature Comparison Terrain and Ground Type

Low Humidity, Low Temperature Comparison

The results for the low humidity, low temperature comparison are shown in Table 29 and Figure 25. The LAMAX results are presented as the mean difference +/- the standard deviations of differences between INM and Nord2000 at the 11 receivers. The terrain and ground type are shown in Figure 26. Table 30 and Table 31 display the difference between each model's baseline and low humidity, low temperature case results.

Nord2000 computes LAMAX 5.2 +/-2.7 dB greater than INM. The maximum difference between the two models for this comparison is 8.5 dB, at the farthest receiver. The differences between each model's baseline and low humidity, low temperature case (Table 30 and Table 31) show a similar trend in INM and Nord2000, with a maximum difference of 0.7 dB.

Table 29. Low Humidity, Low Temperature Results

Receptor Row Fourteen	Distance (m)	INM (dB): Low Humidity, Low Temp	Nord2000 (dB): Low Humidity, Low Temp	Difference (dB)
14_10	0	76.8	78.1	-1.3
14_9	300	72.6	73.8	-1.2
14_8	600	65.8	68.7	-2.9
14_7	900	60.8	64.6	-3.8
14_6	1200	56.8	61.3	-4.5
14_5	1500	53.0	58.6	-5.6
14_4	1800	49.9	56.4	-6.5
14_3	2100	47.1	54.4	-7.2
14_2	2400	44.7	52.5	-7.8
14_1	2700	42.6	50.8	-8.2
14_0	3000	40.7	49.2	-8.5

Table 30. Low Humidity, Low Temperature Comparison, INM Difference from Baseline

Distance (m)	INM (dB): Baseline	INM (dB): Low Humidity, Low Temp	Difference (dB)
0	79.8	76.8	3.0
300	76.0	72.6	3.5
600	69.8	65.8	4.0
900	65.2	60.8	4.4
1200	61.5	56.8	4.7
1500	58.0	53.0	5.0
1800	55.0	49.9	5.2
2100	52.4	47.1	5.3
2400	50.2	44.7	5.4
2700	48.1	42.6	5.5
3000	46.3	40.7	5.6

Table 31. Low Humidity, Low Temperature Comparison, Nord2000 Difference from Baseline

Distance (m)	Nord2000 (dB): Baseline	Nord2000 (dB): Low Humidity, Low Temp	Difference (dB)
0	81.8	78.1	3.7
300	77.9	73.8	4.1
600	72.8	68.7	4.1
900	68.8	64.6	4.3
1200	65.6	61.3	4.4
1500	63.1	58.6	4.5
1800	61.1	56.4	4.7
2100	59.5	54.4	5.1
2400	58.0	52.5	5.5
2700	56.6	50.8	5.8
3000	55.3	49.2	6.1

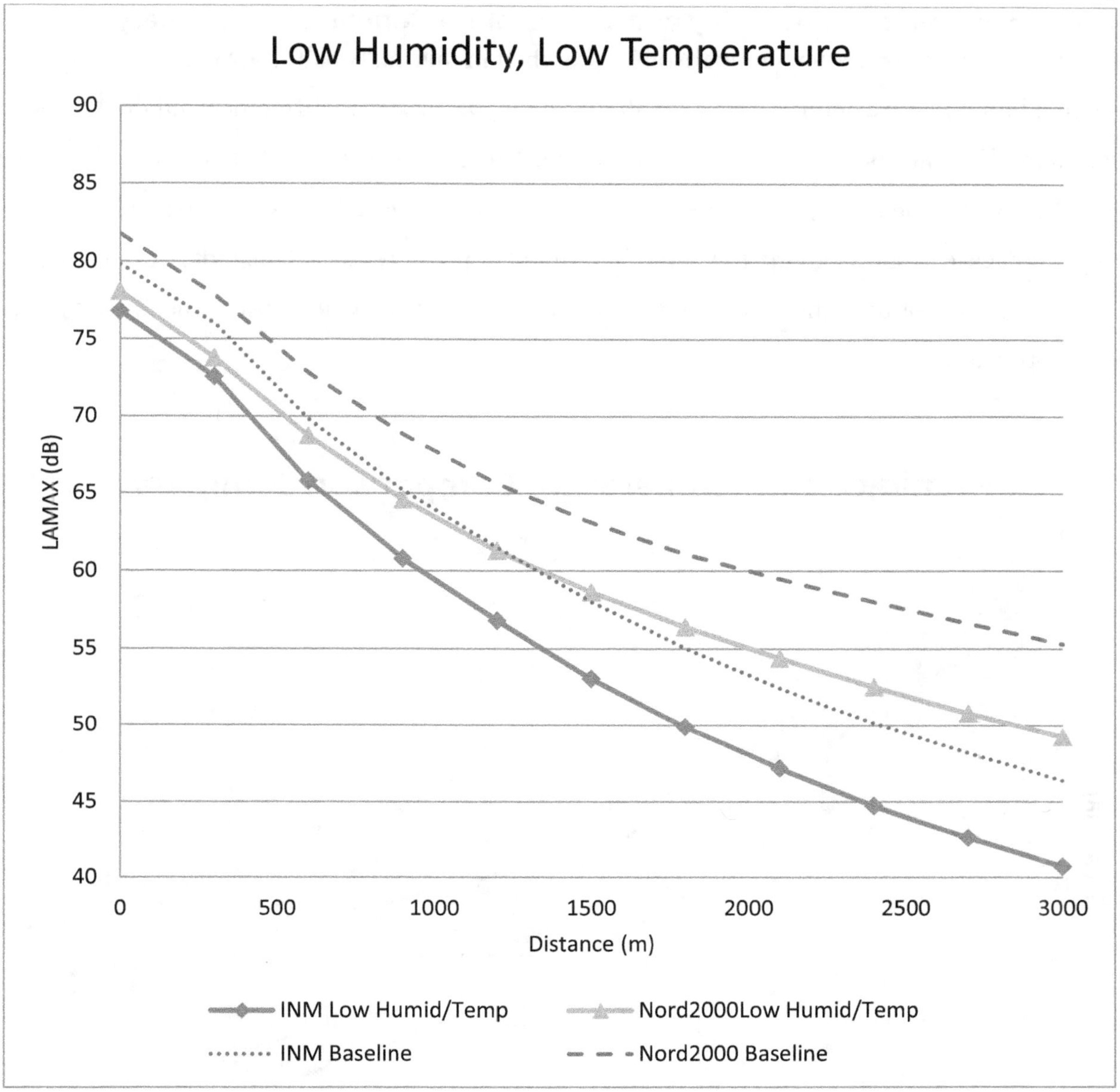

Weather Effects

The Analysis of Modeling Aircraft Noise
with the Nord2000 Noise Model

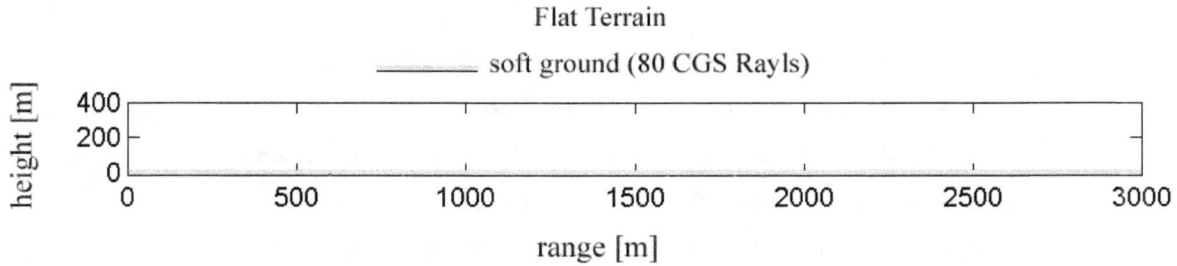

Figure 25. Low Humidity, Low Temperature Results

Figure 26. Low Humidity, Low Temperature Comparison Terrain and Ground Type

8.3.3 Homogeneous Humidity and Temperature Comparison Summary

All humidity and temperature comparisons researched in this analysis follow the same trends. The high humidity, low temperature case shows the highest average difference and the largest maximum difference between INM and Nord2000. The greatest variance between the difference in INM's baseline and comparison case and the difference in Nord2000's baseline and comparison case is shown in the low humidity, high temperature case. This indicates that the atmospheric absorption standards used in INM and Nord2000 vary the most in the low humidity, high temperature case.

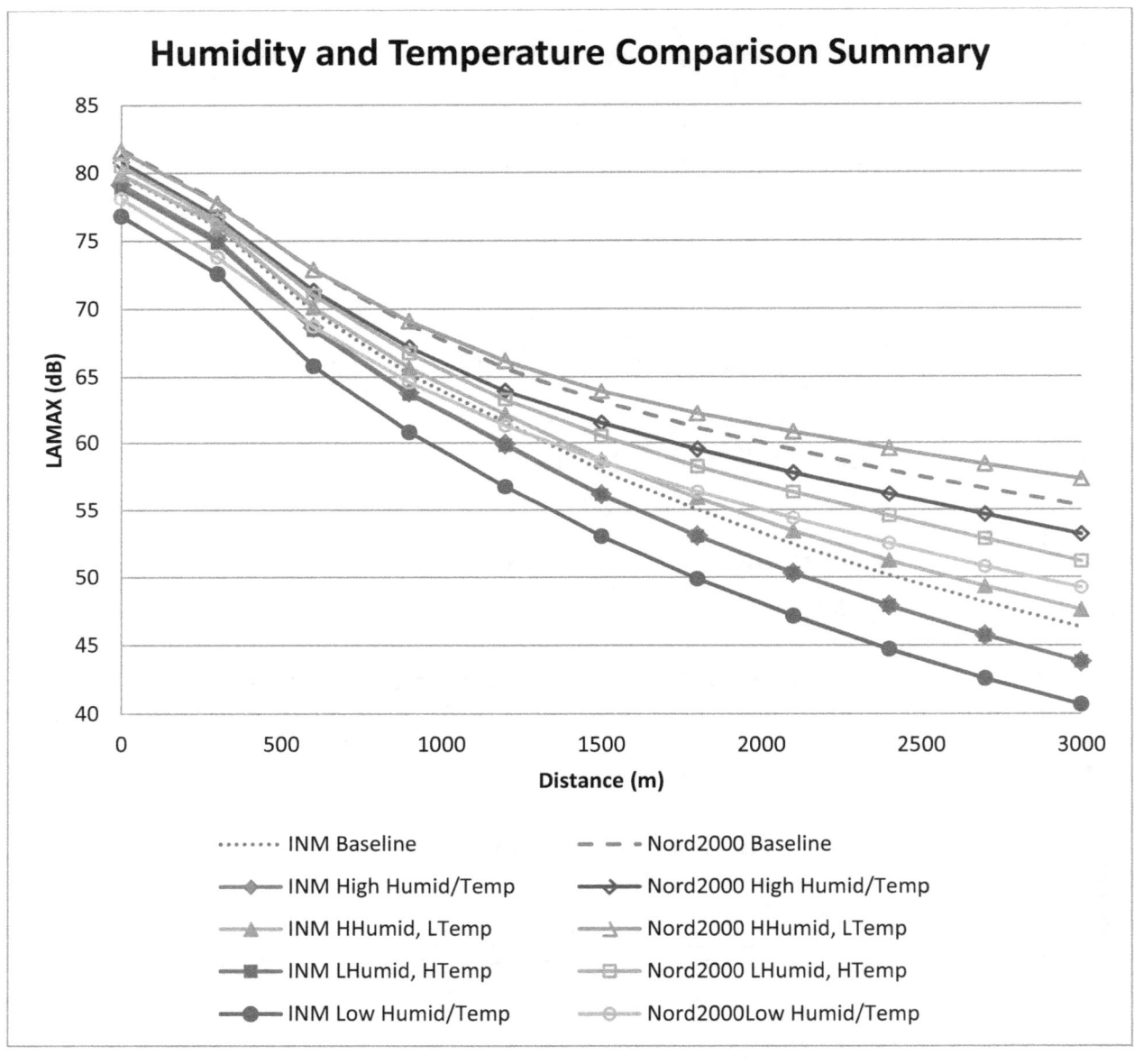

Figure 27. Humidity and Temperature Comparison Summary Graph

8.3.4 Atmospheric Profile Comparison Results
Positive Vertical Temperature Gradient Comparison

A positive vertical temperature gradient, investigated in the section, typically occurs when the ground cools faster than the atmosphere and the temperature of the air increases with height above ground causing downward refraction of rays through the atmosphere. A positive vertical temperature gradient is common during nighttime hours.

The results for the positive temperature gradient comparison are shown in Table 32 and Figure 28. The LAMAX results are presented as the mean difference +/- the standard deviations of differences between INM and Nord2000 at the 11 receivers. The terrain and ground type are shown in Figure 29. Table 33 displays the difference between the Nord2000 positive temperature gradient case and the baseline.

Nord2000 computes LAMAX 5.8 +/-3.0 dB greater than INM. The maximum difference between the two models for this comparison is 10.1 dB, at the farthest receiver. The maximum deviation in Nord2000 from the baseline and this case is 1.2 dB.

Table 32. Positive Temperature Gradient Results

Receptor Row Fourteen	Distance (m)	INM (dB): Baseline	Nord2000 (dB): Positive Temp Gradient	Difference (dB)
14_10	0	79.8	81.8	-1.9
14_9	300	76.0	77.9	-1.8
14_8	600	69.8	73.0	-3.1
14_7	900	65.2	69.1	-3.9
14_6	1200	61.5	66.1	-4.5
14_5	1500	58.0	63.7	-5.7
14_4	1800	55.0	61.7	-6.7
14_3	2100	52.4	60.2	-7.7
14_2	2400	50.2	58.8	-8.7
14_1	2700	48.1	57.6	-9.5
14_0	3000	46.3	56.5	-10.1

Table 33. Positive Temperature Gradient Comparison, Nord2000 Difference from Baseline

Distance (m)	Nord2000 (dB): Baseline	Nord2000 (dB): Positive Temp Gradient	Difference (dB
0	81.8	81.8	0.0
300	77.9	77.9	0.0
600	72.8	73.0	-0.1
900	68.8	69.1	-0.2
1200	65.6	66.1	-0.4
1500	63.1	63.7	-0.6
1800	61.1	61.7	-0.6
2100	59.5	60.2	-0.7
2400	58.0	58.8	-0.8
2700	56.6	57.6	-1.0
3000	55.3	56.5	-1.2

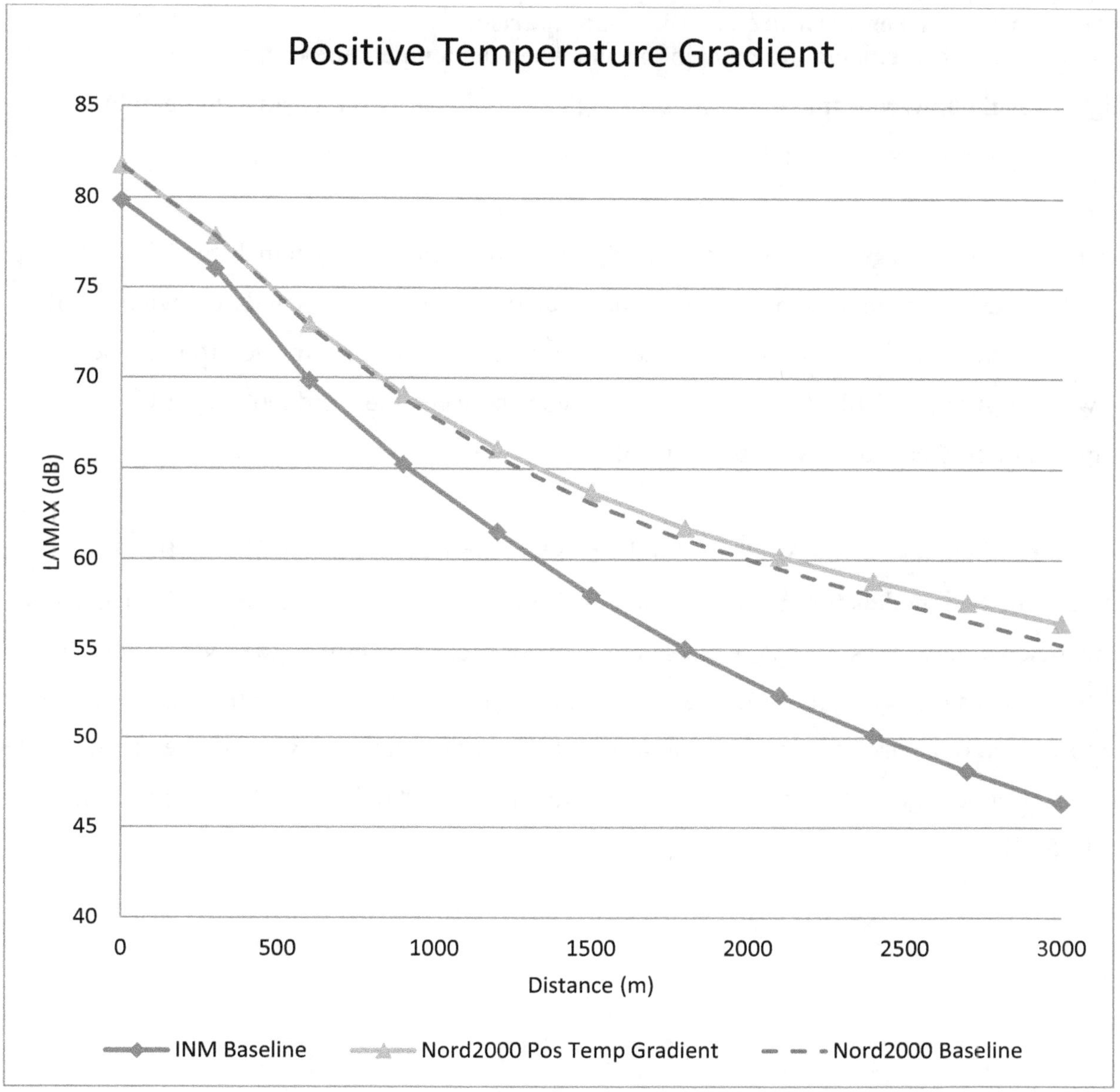

Weather Effects

The Analysis of Modeling Aircraft Noise
with the Nord2000 Noise Model

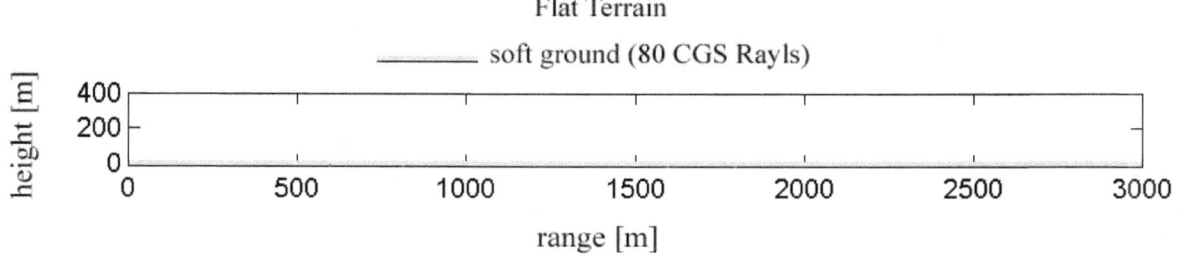

Figure 28. Positive Temperature Gradient Results

Figure 29. Positive Temperature Gradient Comparison Terrain and Ground Type

Negative Vertical Temperature Gradient Comparison

A negative vertical temperature gradient occurs when the temperature of the air decreases with height causing upward refraction of rays through the atmosphere. A negative temperature gradient is common during daytime hours.

The results for the negative temperature gradient comparison are shown in Table 34 and Figure 30. The LAMAX results are presented as the mean difference +/- the standard deviations of differences between INM and Nord2000 at the 11 receivers. The terrain and ground type are shown in Figure 31. Table 35 displays the difference between the Nord2000 negative temperature gradient case and baseline results.

Nord2000 computes LAMAX 4.9 +/- 7.4 dB greater than INM. The maximum difference between the two models for this comparison is 12.8 dB, at the farthest receiver. This is the only comparison where the Nord2000 results are not exclusively greater than the INM results. In fact, the two models cross just after receiver 14_6. The drop in level of Nord2000 is likely caused by the formation of a shadow zone. In conditions of upward refraction, levels are greatly attenuated in shadow zone regions[12]. The maximum deviation in Nord2000 from the baseline and this case is 21.75 dB.

Table 34. Negative Temperature Gradient Results

Receptor Row Fourteen	Distance (m)	INM (dB): Baseline	Nord2000 (dB): Negative Temp Gradient	Difference (dB)
14_10	0	79.8	81.8	-1.9
14_9	300	76.0	78.0	-1.9
14_8	600	69.8	72.7	-2.8
14_7	900	65.2	68.6	-3.4
14_6	1200	61.5	65.2	-3.7
14_5	1500	58.0	47.0	11.0
14_4	1800	55.0	43.5	11.6
14_3	2100	52.4	41.2	11.2
14_2	2400	50.2	40.0	10.1
14_1	2700	48.1	36.8	11.4
14_0	3000	46.3	33.5	12.8

Table 35. Negative Temperature Gradient Comparison, Nord2000 Difference from Baseline

Distance (m)	Nord2000 (dB): Baseline	Nord2000 (dB): Negative Temp Gradient	Difference (dB)
0	81.8	81.8	0.00
300	77.9	78.0	-0.08
600	72.8	72.7	0.16
900	68.8	68.6	0.22
1200	65.6	65.2	0.46
1500	63.1	47.0	16.12
1800	61.1	43.5	17.64
2100	59.5	41.2	18.29
2400	58.0	40.0	17.98
2700	56.6	36.8	19.83
3000	55.3	33.5	21.75

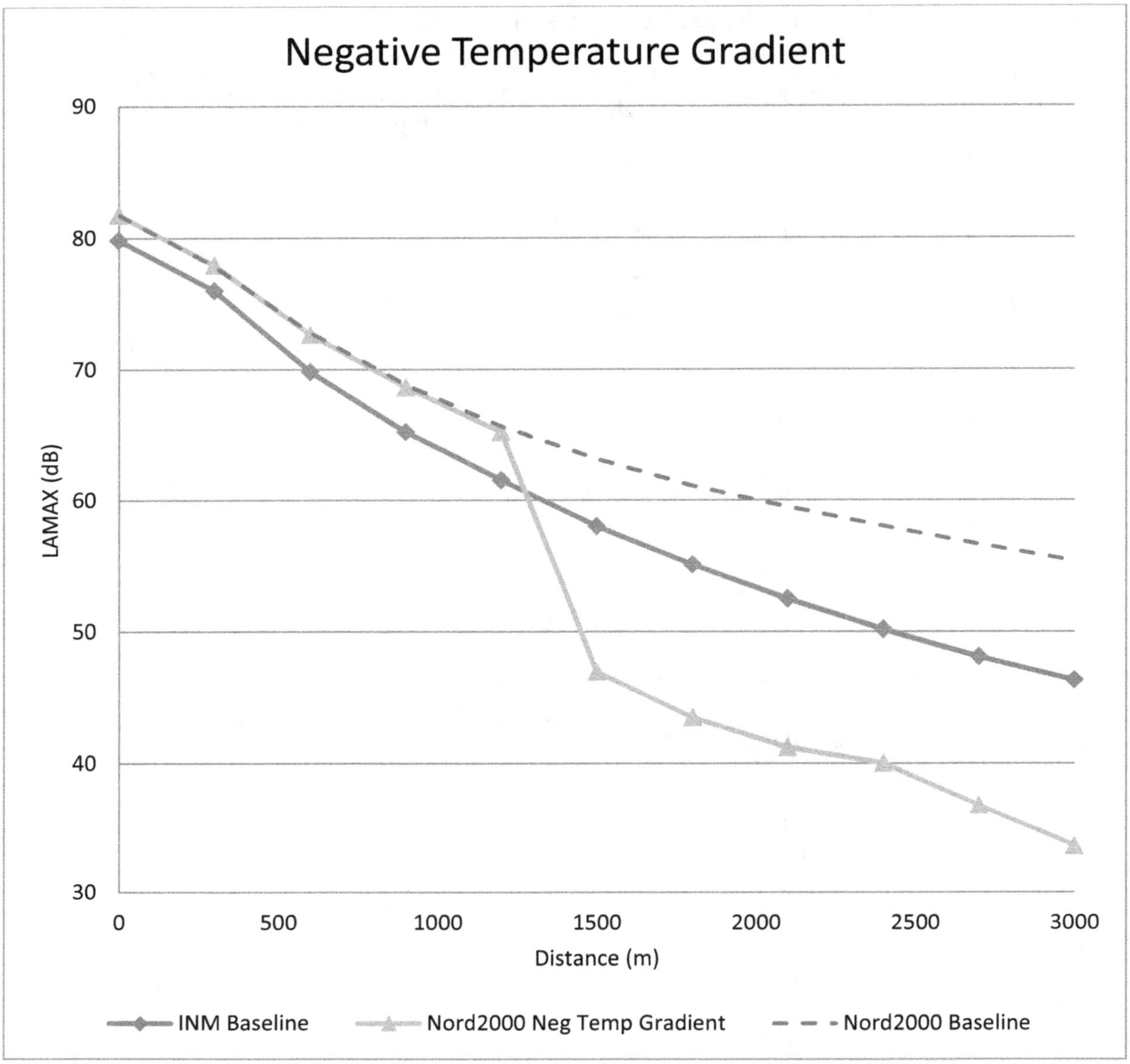

Weather Effects

The Analysis of Modeling Aircraft Noise with the Nord2000 Noise Model

Figure 30. Negative Temperature Gradient Results

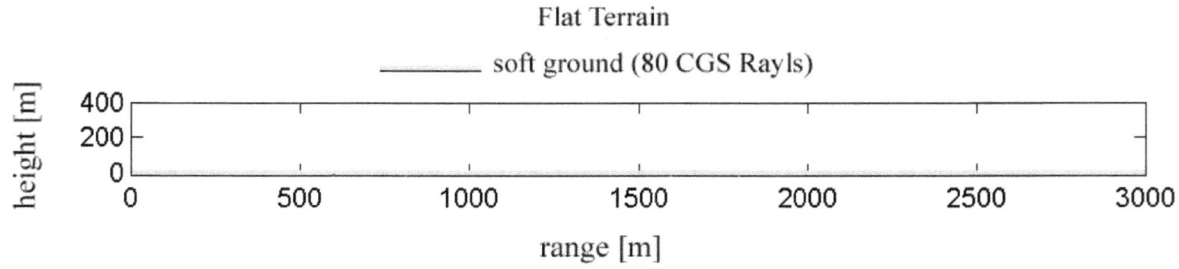

Figure 31. Negative Temperature Gradient Comparison Terrain and Ground Type

8.3.5 Temperature Gradient Comparison Summary

Figure 32 shows the effect of both temperature gradient comparisons researched, compared with the baseline conditions. The difference between the Nord2000 baseline and positive temperature gradient cases is minimal, however the difference between the Nord2000 baseline and negative temperature gradient cases is significant for distances beyond 1,200 m.

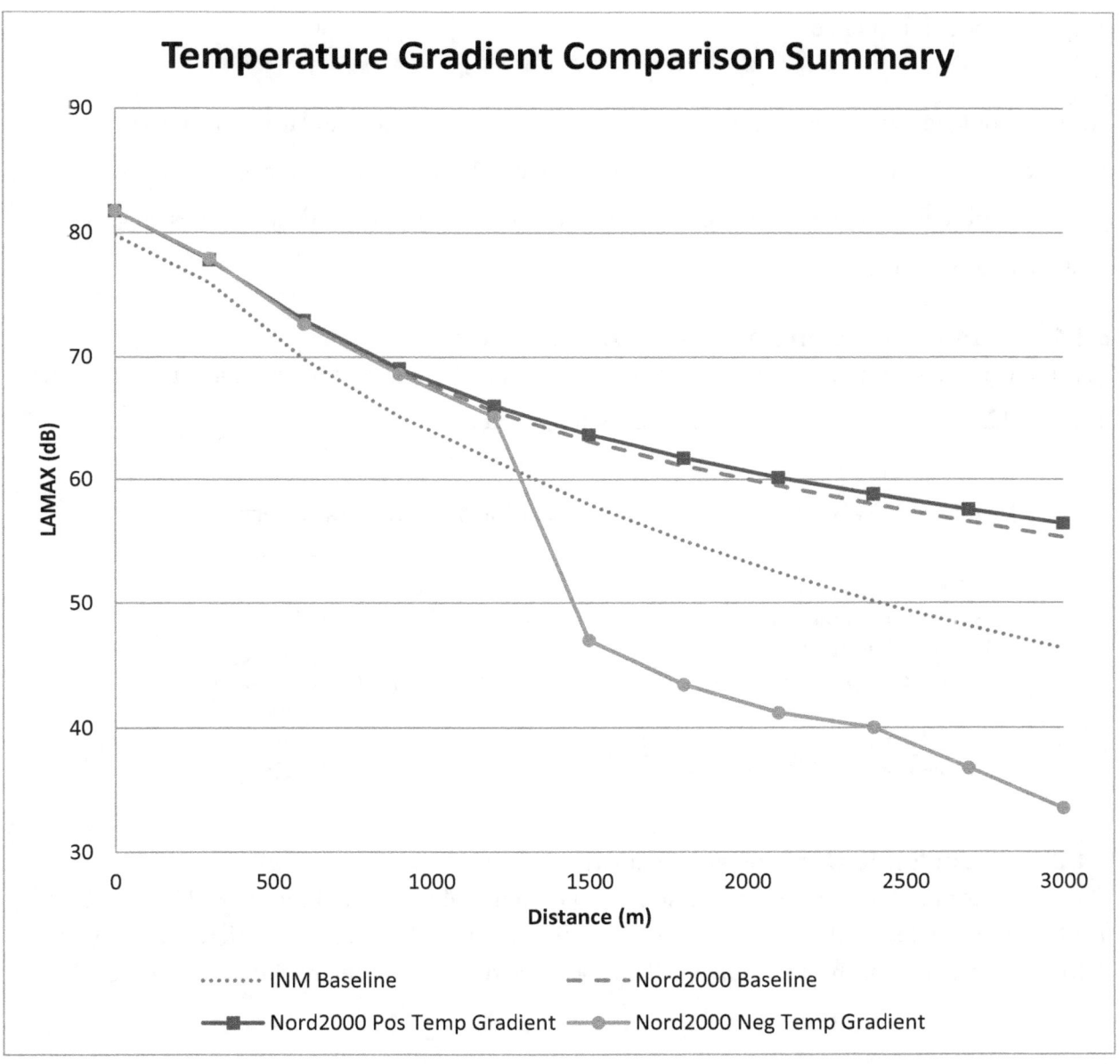

Figure 32. Temperature Gradient Comparison Summary Graph

8.4 Turbulence Comparisons

Nord2000 allows for user input of atmospheric turbulence parameters for wind and temperature to calculate the effect of atmospheric turbulence. The effect of turbulence is included as a reduction in coherence between different rays. An additional effect of turbulence in shadow zones is only applied if the line of sight between the source and receiver is broken[7]. Shadow zones could be created by terrain features or an upward refracting atmosphere[8]. Conversely, INM does not model turbulence.

Strong turbulence was modeled in Nord2000 with flat terrain and standard atmospheric conditions and compared against the INM baseline conditions in order to capture the most extreme differences between the models. The effect of turbulence in shadow zones is investigated further in Section 9.

8.4.1 Turbulence Comparison Input Parameters

Turbulence-related input parameters are shown in Table 36. Parameters that have been changed from the baseline conditions are identified by italicized text.

Table 36. Turbulence Comparison Input Parameters

	INM	Nord2000
Temperature	15°C	15°C
Temperature Gradient	0	0
Relative Humidity	70%	70%
Turbulence	N/A	Wind: $0.000680697 m^{4/3} s^{-2}$
		Temperature: $0.049818254 K^2 m^{-2/3}$
Terrain	Flat	Flat
Ground Type	Soft, 80 CGS Rayls	Soft, 80 CGS Rayls

8.4.2 Turbulence Comparison Results

The results for the turbulence comparison are shown in Table 37 and Figure 33. The LAMAX results are presented as the mean difference +/- the standard deviations of differences between INM and Nord2000 at the 11 receivers. The terrain and ground type are shown in Figure 34.

Nord2000 computes LAMAX 5.3 +/- 2.6 dB greater than INM. The maximum difference between the two models is 8.9 dB, at the farthest receiver. The maximum deviation in Nord2000 between the baseline and this case is only 0.1 dB indicating that turbulence is not a major contributor to the differences between INM and Nord2000.

A brief proof of concept investigation into turbulence effects over longer range propagation where grazing angles of the ground-reflected ray are very shallow indicated that turbulence could have a significant effect. However, turbulence had a much smaller effect for any condition with the source-receiver geometries used in this report.

Table 37. Turbulence Comparison Results

Receptor Row Fourteen	Distance (m)	INM (dB): Baseline	Nord2000 (dB): Strong Turbulence	Difference (dB)
14_10	0	79.8	81.8	-1.9
14_9	300	76.0	77.9	-1.3
14_8	600	69.8	72.8	-3.0
14_7	900	65.2	68.9	-3.7
14_6	1200	61.5	65.7	-4.1
14_5	1500	58.0	63.1	-5.1
14_4	1800	55.0	61.2	-6.1
14_3	2100	52.4	59.5	-7.0
14_2	2400	50.2	58.0	-7.8
14_1	2700	48.1	56.6	-8.4
14_0	3000	46.3	55.2	-8.9

Table 38. Turbulence Comparison, Nord2000 Difference from Baseline

Distance (m)	Nord2000 (dB): Baseline	Nord2000 (dB): Strong Turbulence	Difference (dB)
0	81.8	81.8	0.0
300	77.9	77.9	0.0
600	72.8	72.8	0.0
900	68.8	68.9	0.0
1200	65.6	65.7	0.0
1500	63.1	63.1	0.0
1800	61.1	61.2	-0.1
2100	59.5	59.5	0.0
2400	58.0	58.0	0.0
2700	56.6	56.6	0.1
3000	55.3	55.2	0.1

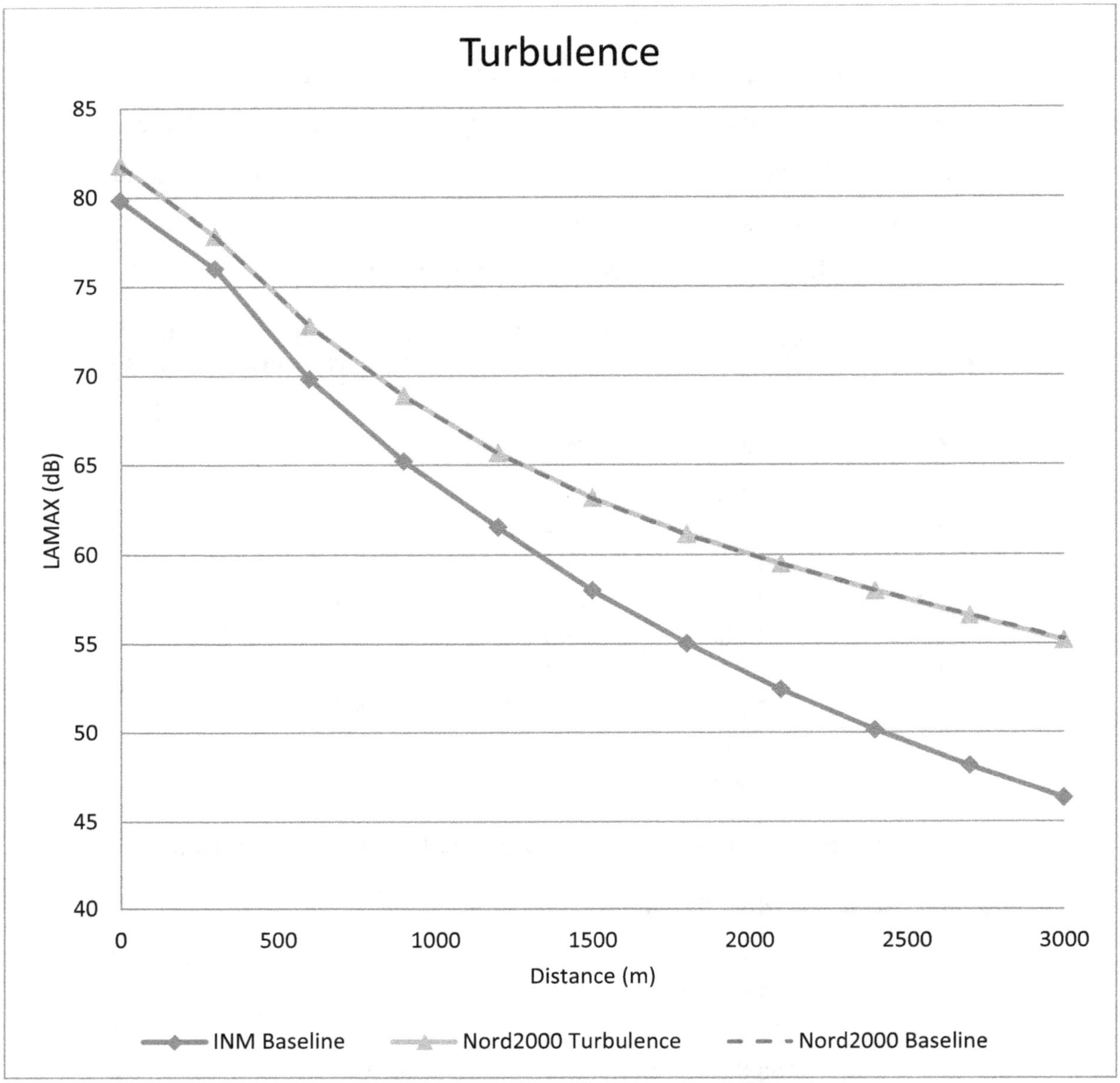

Turbulence

Figure 33. Turbulence Comparison Results

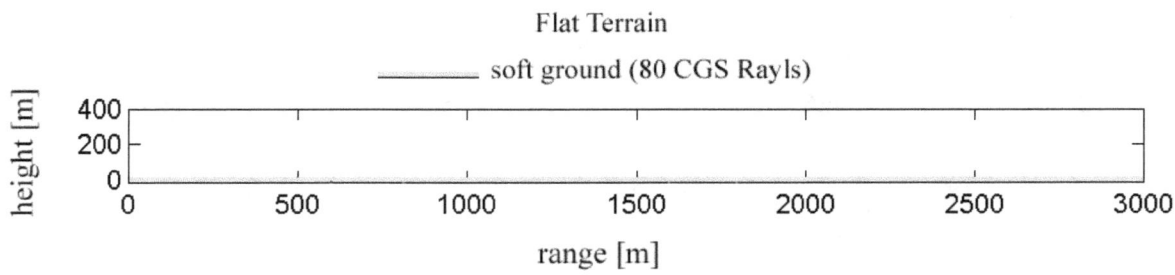

Figure 34. Turbulence Comparison Terrain and Ground Type

9 MULTIPLE PARAMETER VARIATION COMPARISONS

In order to investigate model performance for more realistic scenarios, multiple parameter variation comparisons were conducted for the following conditions:

- Mixed Ground Type - 1
- Mixed Ground Type - 2
- Mixed Ground Type - 3
- Hill Terrain, Mixed Ground Type - Configuration 3
- Hill Terrain, Mixed Ground Type - Configuration 3, Turbulence
- Hill Terrain, Positive Temperature Gradient

9.1 Multiple Parameter Variation Input Parameters

INM does not allow ground type input so all mixed ground type multiple parameter variation comparisons done in Nord2000 are compared to the INM baseline conditions. INM allows for terrain input; therefore Nord2000 comparisons with multiple parameter variations that are run with hill terrain are compared to the INM baseline with hill terrain and baseline conditions. The input parameters for Nord2000 are shown in Table 39 and Table 40.

Table 39. Nord2000 Multiple Parameter Variation Input Parameters-1

	Mixed Ground Type-1	Mixed Ground Type-2	Mixed Ground Type-3
Temperature	15°C	15°C	15°C
Temperature Gradient	0	0	0
Relative Humidity	70%	70%	70%
Turbulence	0	0	0
Terrain	Flat	Flat	Flat
Ground Type	0-600 m: 20000 CGS Rayls	0-600 m: 80 CGS Rayls	0-600 m: 20000 CGS Rayls
	600-3000 m: 80 CGS Rayls	600-3000 m: 20000 CGS Rayls	600-1800 m: 80 CGS Rayls
			1800-3000 m: 20000 CGS Rayls

Table 40. Nord2000 Multiple Parameter Variation Input Parameters-2

	Hill Terrain, MG-3	Hill Terrain, MG-3, Turbulence	Hill Terrain, Pos Temp Gradient
Temperature	15°C	15°C	15°C
Temperature Gradient	0	0	0.2032 °C/m
Relative Humidity	70%	70%	70%
Turbulence	0	0	0
Terrain	*Hill*	*Hill*	*Hill*
Terrain Feature Height	*70 m*	*70 m*	*70 m*
Ground Type	*0-600 m: 20000 CGS Rayls*	*0-600 m: 20000 CGS Rayls*	Soft, 80 CGS Rayls
	600-1800 m: 80 CGS Rayls	*600-1800 m: 80 CGS Rayls*	
	1800-3000 m: 20000 CGS Rayls	*1800-3000 m: 20000 CGS Rayls*	

9.2 Multiple Parameter Variation Results – Mixed Ground Type

Mixed Ground Type – 1 Comparison

The results for the mixed ground type -1 (0-600 m hard ground, 600-3000 m soft ground) comparison are shown in Table 41 and Figure 35. The LAMAX results are presented as the mean difference +/- the standard deviations of differences between INM and Nord2000 at the 11 receivers. The terrain and ground type are shown in Figure 36. The vertical dotted lines indicate ranges at which transitions in propagation conditions occur. Table 42 displays the difference between the Nord2000 mixed ground type-1 case and baseline results.

Nord2000 computes LAMAX 5.8 +/- 2.4 dB greater than INM. The maximum difference between the two models for this comparison is 9.0 dB, at the farthest receiver. Note that the differences between the Nord2000 baseline and this case appear over the distances 0-600 m, where the ground type differs from the baseline.

Table 41. Mixed Ground Type-1 Results

Receptor Row Fourteen	Distance (m)	INM (dB): Baseline	Nord2000 (dB): Mixed Ground 1	Difference (dB)
14_10	0	79.8	83.7	-3.8
14_9	300	76.0	79.8	-3.8
14_8	600	69.8	72.9	-3.1
14_7	900	65.2	68.8	-3.6
14_6	1200	61.5	65.6	-4.1
14_5	1500	58.0	63.1	-5.1
14_4	1800	55.0	61.1	-6.1
14_3	2100	52.4	59.5	-7.1
14_2	2400	50.2	58.0	-7.9
14_1	2700	48.1	56.6	-8.5
14_0	3000	46.3	55.3	-9.0

Table 42. Mixed Ground Type-1 Comparison, Nord2000 Difference from Baseline

Distance (m)	Nord2000 (dB): Baseline	Nord2000 (dB): Mixed Ground 1	Difference (dB)
0	81.8	83.7	-1.9
300	77.9	79.8	-1.9
600	72.8	72.9	-0.1
900	68.8	68.8	0.0
1200	65.6	65.6	0.0
1500	63.1	63.1	0.0
1800	61.1	61.1	0.0
2100	59.5	59.5	0.0
2400	58.0	58.0	0.0
2700	56.6	56.6	0.0
3000	55.3	55.3	0.0

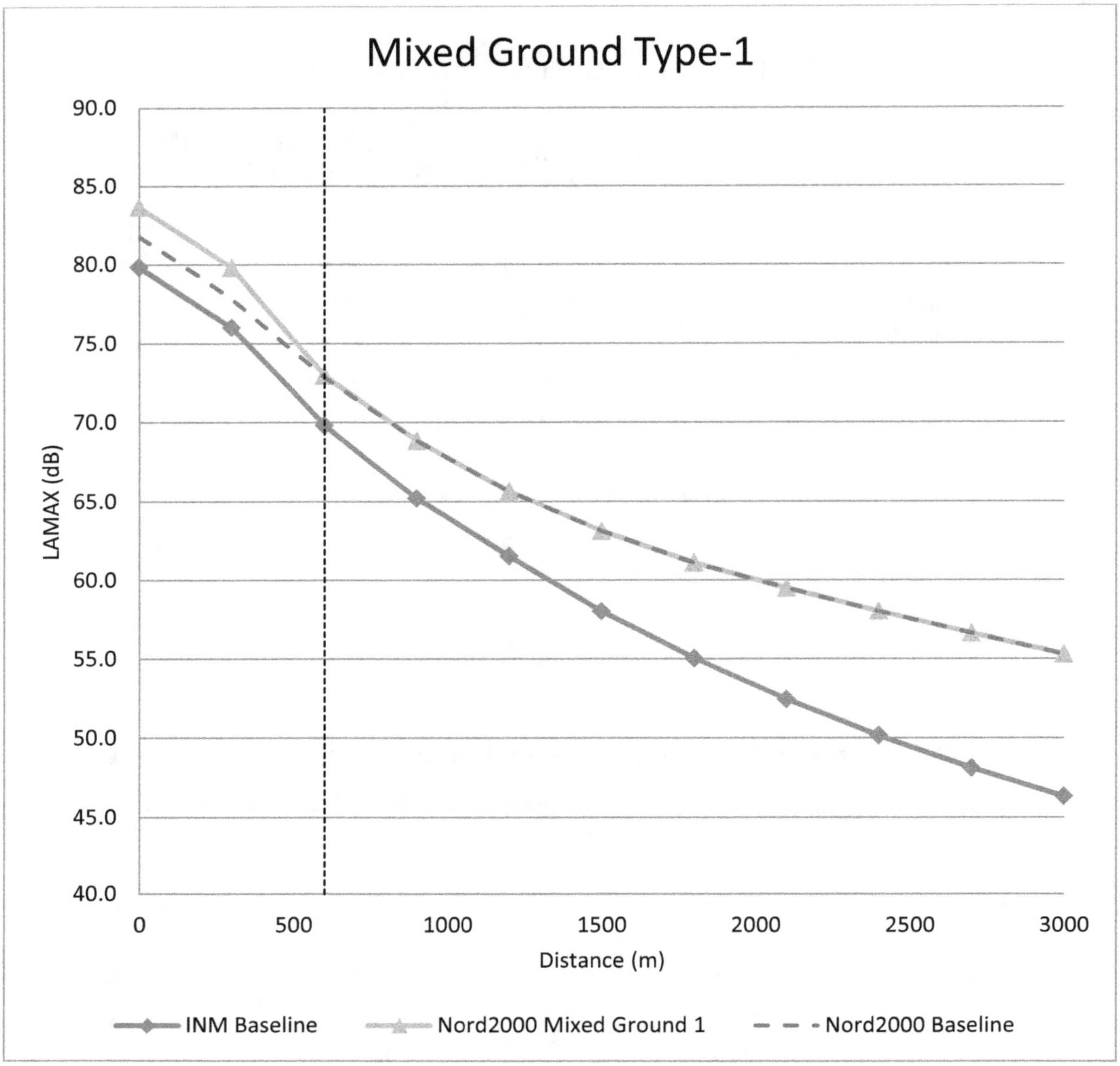

Multiple Parameter Variation Comparisons

The Analysis of Modeling Aircraft Noise with the Nord2000 Noise Model

Figure 35. Mixed Ground Type-1 Results

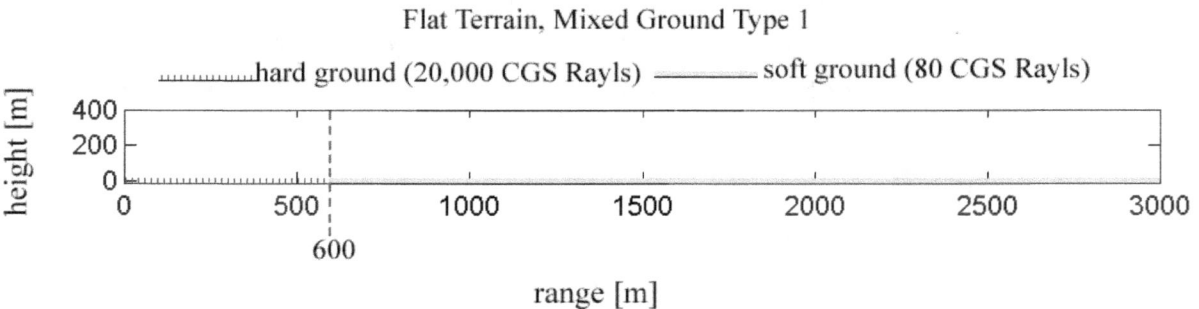

Figure 36. Mixed Ground Type-1 Terrain and Ground Type

Mixed Ground Type – 2 Comparison

The results for the mixed ground type -2 (0-600 m soft ground, 600-2000 m hard ground) comparison are shown in Table 43 and Figure 37. The LAMAX results are presented as the mean difference +/- the standard deviations of differences between INM and Nord2000 at the 11 receivers. The terrain and ground type are shown in Figure 38. The vertical dotted lines indicate ranges at which transitions in propagation conditions occur. Table 44 displays the difference between the Nord2000 mixed ground type-2 case and baseline results.

Nord2000 computes LAMAX 6.6 +/- 3.0 dB greater than INM. The maximum difference between the two models for this comparison is 10.9 dB, at the farthest receiver. Note that the differences between the Nord2000 baseline and this case appear over the distances 600-3000 m, where the ground type differs from the baseline.

Table 43. Mixed Ground Type-2 Results

Receptor Row Fourteen	Distance (m)	INM (dB): Baseline	Nord2000 (dB): Mixed Ground 2	Difference (dB)
14_10	0	79.8	81.8	-1.9
14_9	300	76.0	77.9	-1.8
14_8	600	69.8	74.6	-4.7
14_7	900	65.2	70.9	-5.7
14_6	1200	61.5	67.6	-6.0
14_5	1500	58.0	64.9	-6.9
14_4	1800	55.0	62.6	-7.6
14_3	2100	52.4	60.7	-8.2
14_2	2400	50.2	59.2	-9.0
14_1	2700	48.1	58.1	-10.0
14_0	3000	46.3	57.3	-10.9

Table 44. Mixed Ground Type-2 Comparison, Nord2000 Difference from Baseline

Distance (m)	Nord2000 (dB): Baseline	Nord2000 (dB): Mixed Ground 2	Difference (dB)
0	81.8	81.8	0.0
300	77.9	77.9	0.0
600	72.8	74.6	-1.7
900	68.8	70.9	-2.0
1200	65.6	67.6	-1.9
1500	63.1	64.9	-1.8
1800	61.1	62.6	-1.5
2100	59.5	60.7	-1.2
2400	58.0	59.2	-1.2
2700	56.6	58.1	-1.5
3000	55.3	57.3	-2.0

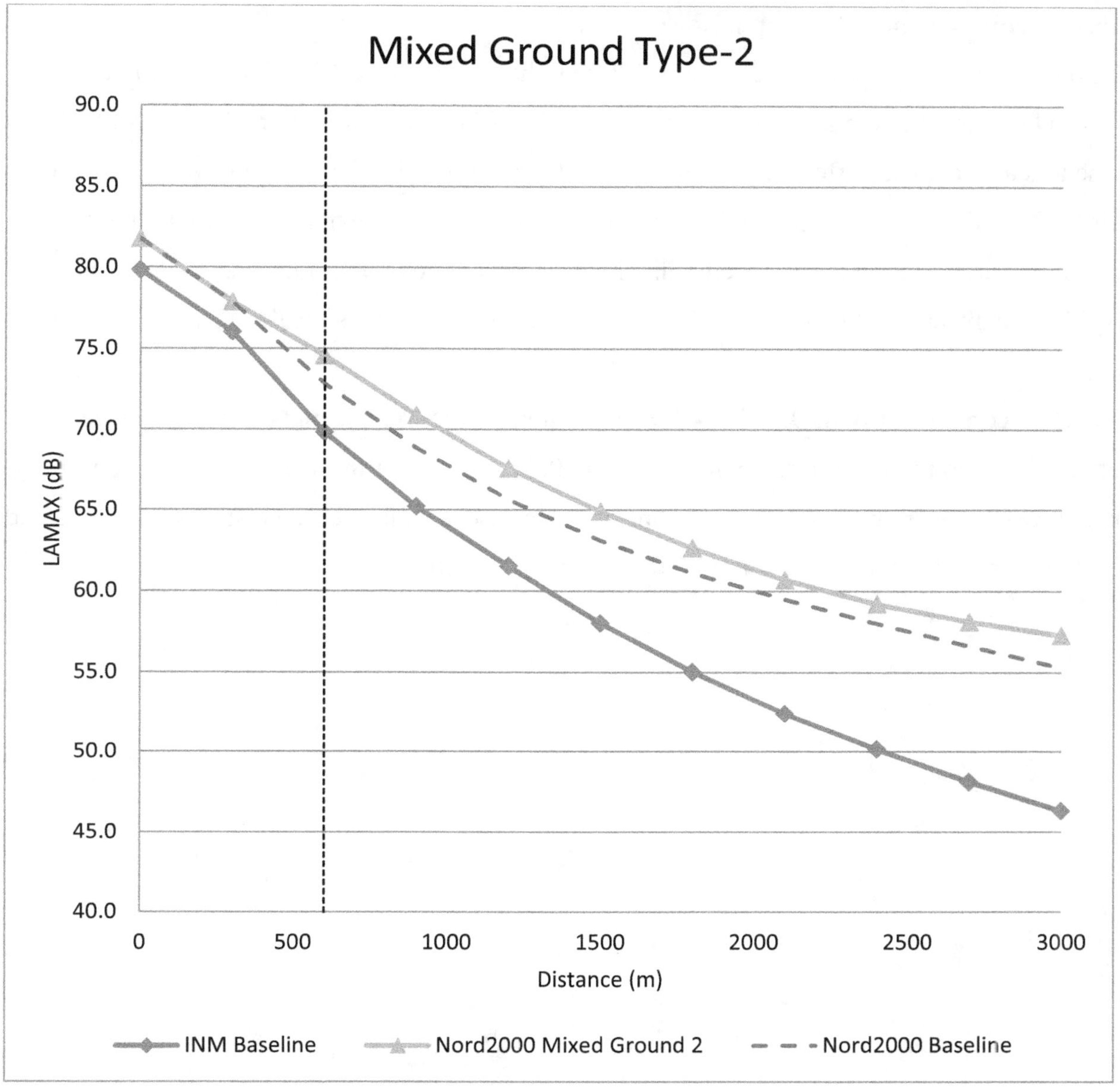

Multiple Parameter Variation Comparisons

The Analysis of Modeling Aircraft Noise with the Nord2000 Noise Model

Figure 37. Mixed Ground Type-2 Results

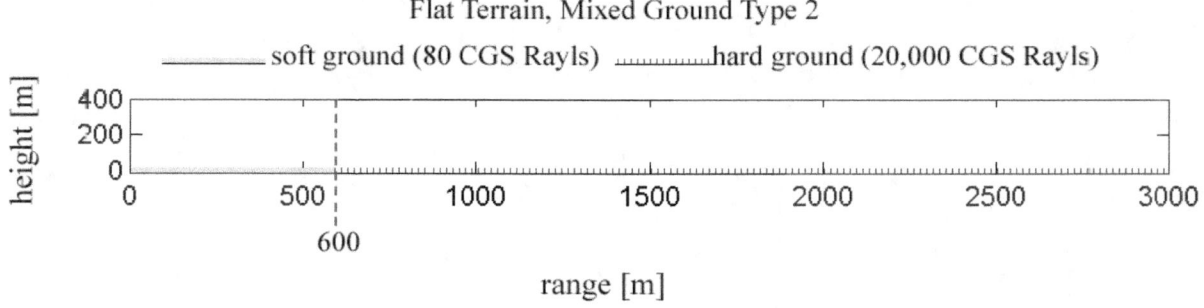

Figure 38. Flat Terrain, Mixed Ground Type-2

Mixed Ground Type – 3 Comparison

The results for the mixed ground type -3 (0-600 m hard ground, 600-1800 m soft ground, 1800-3000 m hard ground) comparison are shown in Table 45 and Figure 39. The LAMAX results are presented as the mean difference +/- the standard deviations of differences between INM and Nord2000 at the 11 receivers. The terrain and ground type are shown in Figure 40. The vertical dotted lines indicate ranges at which transitions in propagation conditions occur. Table 46 displays the difference between the Nord2000 mixed ground type-3 case and baseline results.

Nord2000 computes LAMAX 6.2 +/- 2.9 dB greater than INM. The maximum difference between the two models for this comparison is 10.9 dB, at the farthest receiver. Note that the differences between the Nord2000 baseline and this case appear over the distances 0-600 m and 1800-3000 m, where the ground type differs from the baseline.

Table 45. Mixed Ground Type-3 Results

Receptor Row Fourteen	Distance (m)	INM (dB): Baseline	Nord2000 (dB): Mixed Ground 3	Difference (dB)
14_10	0	79.8	83.7	-3.8
14_9	300	76.0	79.8	-3.8
14_8	600	69.8	72.9	-3.1
14_7	900	65.2	68.8	-3.6
14_6	1200	61.5	65.6	-4.1
14_5	1500	58.0	63.1	-5.1
14_4	1800	55.0	61.8	-6.8
14_3	2100	52.4	60.7	-8.2
14_2	2400	50.2	59.2	-9.0
14_1	2700	48.1	58.1	-10.0
14_0	3000	46.3	57.3	-10.9

Table 46. Mixed Ground Type-3 Comparison, Nord2000 Difference from Baseline

Distance (m)	Nord2000 (dB): Baseline	Nord2000 (dB): Mixed Ground 3	Difference (dB)
0	81.8	83.7	-1.9
300	77.9	79.8	-1.9
600	72.8	72.9	-0.1
900	68.8	68.8	0.0
1200	65.6	65.6	0.0
1500	63.1	63.1	0.0
1800	61.1	61.8	-0.7
2100	59.5	60.7	-1.2
2400	58.0	59.2	-1.2
2700	56.6	58.1	-1.5
3000	55.3	57.3	-2.0

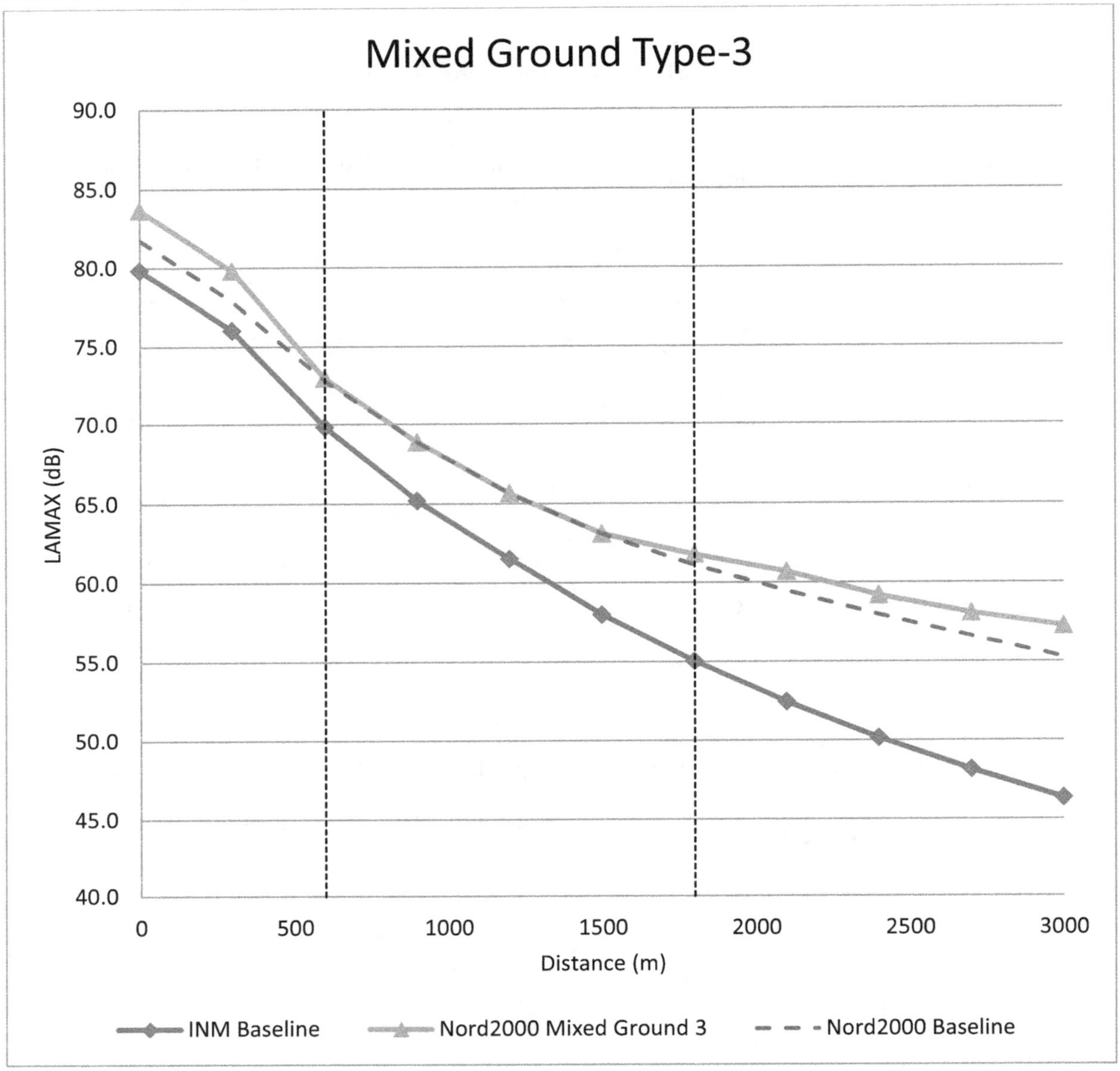

Mixed Ground Type-3

Figure 39. Mixed Ground Type-3 Results

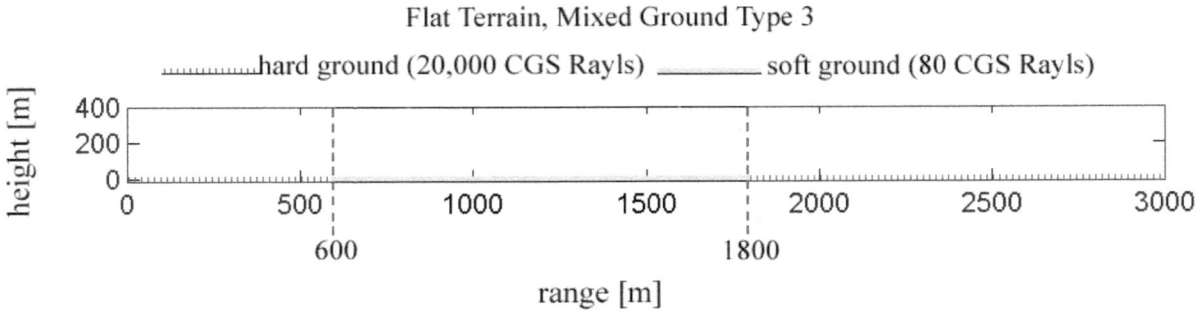

Figure 40. Flat Terrain, Mixed Ground Type-3

9.3 Mixed Ground Type Comparison Summary

All mixed ground type comparisons show expected trends; the results increase from the baseline when the ground type is changed to hard ground, and the results match the baseline results when the ground type is soft ground. The maximum variation from the baseline when changing ground type was 2 dB.

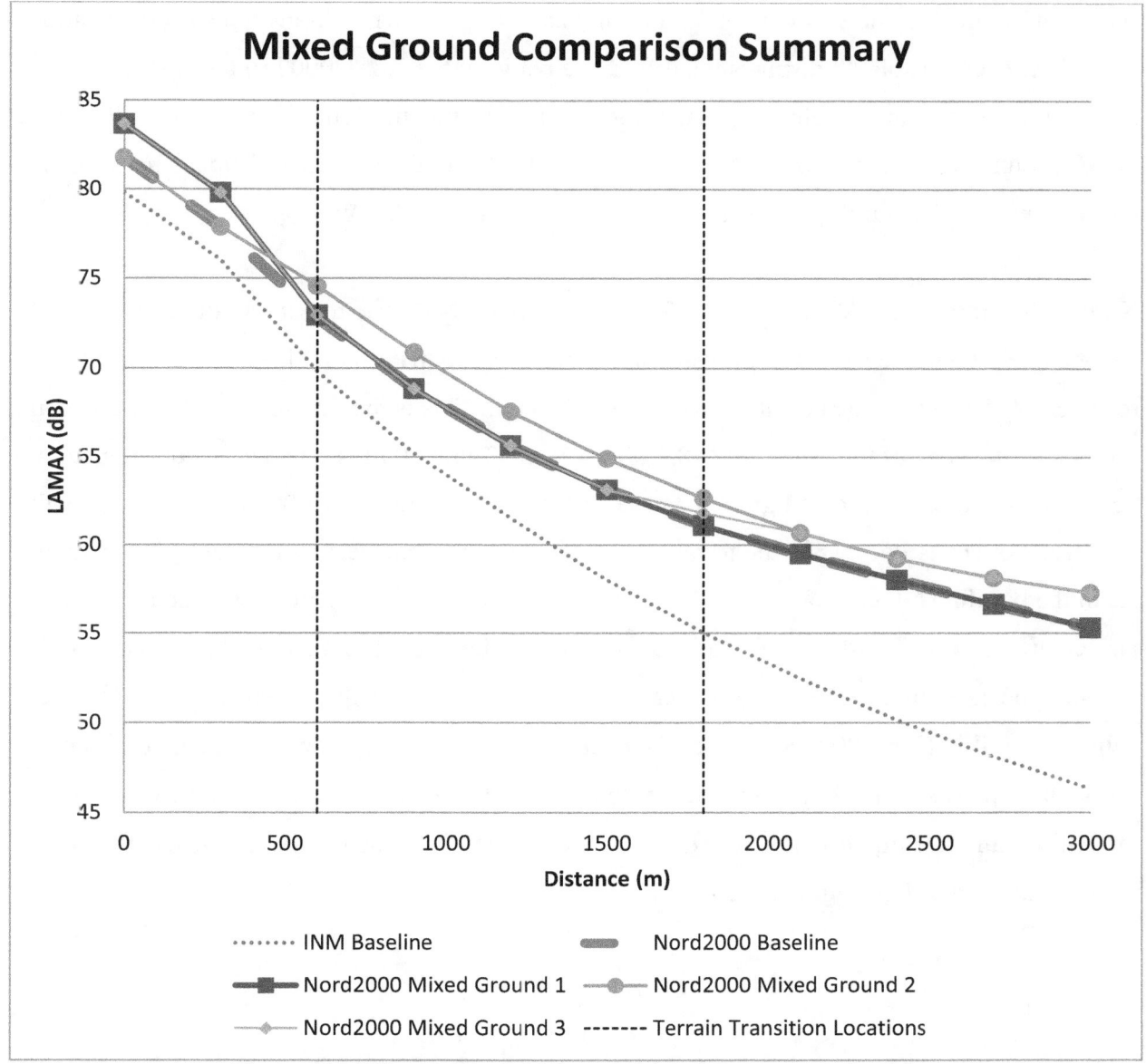

Figure 41. Mixed Ground Comparison Summary Graph

9.4 Multiple Parameter Variation Results – Hill Terrain Type
Hill Terrain, Mixed Ground Type – 3 Comparison

The results for the hill terrain, mixed ground type -3 comparison are shown in Table 47 and Figure 42. The LAMAX results are presented as the mean difference +/- the standard deviations of differences between INM and Nord2000 at the 11 receivers. The terrain and ground type are shown in Figure 43. The vertical dotted lines indicate ranges at which transitions in propagation conditions occur. Table 48 displays the difference between the Nord2000 hill terrain, mixed ground type-3 case and baseline results. Table 49 displays the difference between the hill terrain, mixed ground type-3 case and the hill terrain case. Table 50 displays the difference between the hill terrain, mixed ground type-3 case and the mixed ground type-3 case.

Nord2000 computes LAMAX 7.6 +/- 4.5 dB greater than INM. The maximum difference between the two models for this comparison is 18.4 dB. The maximum difference occurs at receiver 14_5 which is located at approximately 1500 m. This is the first receiver to report sound levels past the peak of the hill. Table 48, Table 49, and Table 50 illustrate that the differences between the hill terrain, mixed ground 3 case and the baseline in Nord2000 are dominated by the effective flow resistivity when the terrain is flat and by the terrain feature between 600-1800 m. As in the baseline hill comparison with soft ground, Section 7.2.2, the increase in sound level, particularly at receiver 14_5 (1500 m), was not expected in the Nord2000 results for the mixed ground type case. In the Mixed Ground case, even louder noise levels were observed at receivers behind the hill in Nord2000, because of the decreased ground absorption over the hard ground portions of the case. It is hypothesized that this increase in noise would be even greater behind the Hill for an all Hard Ground case. Further investigation is needed to determine the source of the increase with hill terrain feature present.

Table 47. Hill Terrain, Mixed Ground Type-3 Results

Receptor Row Fourteen	Distance (m)	INM (dB): Hill	Nord2000(dB): Hill, Mixed ground 3	Difference (dB)
14_10	0	79.8	83.7	-3.8
14_9	300	75.7	79.8	-4.1
14_8	600	69.8	72.9	-3.1
14_7	900	65.1	68.9	-3.8
14_6	1200	60.7	65.9	-5.3
14_5	1500	49.4	67.8	-18.4
14_4	1800	55.0	62.0	-7.0
14_3	2100	52.4	60.8	-8.4
14_2	2400	50.2	59.3	-9.2
14_1	2700	48.1	58.2	-10.1
14_0	3000	46.3	57.3	-11.0

Table 48. Hill Terrain, Mixed Ground Type-3 Comparison, Nord2000 Difference from Baseline

Distance (m)	Nord2000 (dB): Baseline	Nord2000 (dB): Hill, Mixed ground 3	Difference (dB)
0	81.8	83.7	-1.9
300	77.9	79.8	-1.9
600	72.8	72.9	-0.1
900	68.8	68.9	0.0
1200	65.6	65.9	-0.3
1500	63.1	67.8	-4.7
1800	61.1	62.0	-0.9
2100	59.5	60.8	-1.3
2400	58.0	59.3	-1.3
2700	56.6	58.2	-1.6
3000	55.3	57.3	-2.0

Table 49. Difference between the Hill Terrain, Mixed Ground Type-3 Case and Hill Terrain Case

Distance (m)	Nord2000(dB): Hill, Mixed ground 3	Nord2000 (dB): Hill	Difference (dB)
0	83.7	81.8	1.9
300	79.8	77.9	1.9
600	72.9	72.8	0.1
900	68.9	68.9	0.0
1200	65.9	65.9	0.0
1500	67.8	65.0	2.8
1800	62.0	61.4	0.6
2100	60.8	59.7	1.1
2400	59.3	58.1	1.2
2700	58.2	56.7	1.5
3000	57.3	55.3	2.0

Table 50. Difference between the Hill Terrain, Mixed Ground 3 Case and the Mixed Ground 3 Case

Distance (m)	Nord2000(dB): Hill, Mixed ground 3	Nord2000 (dB): Mixed Ground 3	Difference (dB)
0	83.7	83.7	0.0
300	79.8	79.8	0.0
600	72.9	72.9	0.0
900	68.9	68.8	0.1
1200	65.9	65.6	0.3
1500	67.8	63.1	4.7
1800	62.0	61.8	0.2
2100	60.8	60.7	0.1
2400	59.3	59.2	0.1
2700	58.2	58.1	0.1
3000	57.3	57.3	0.0

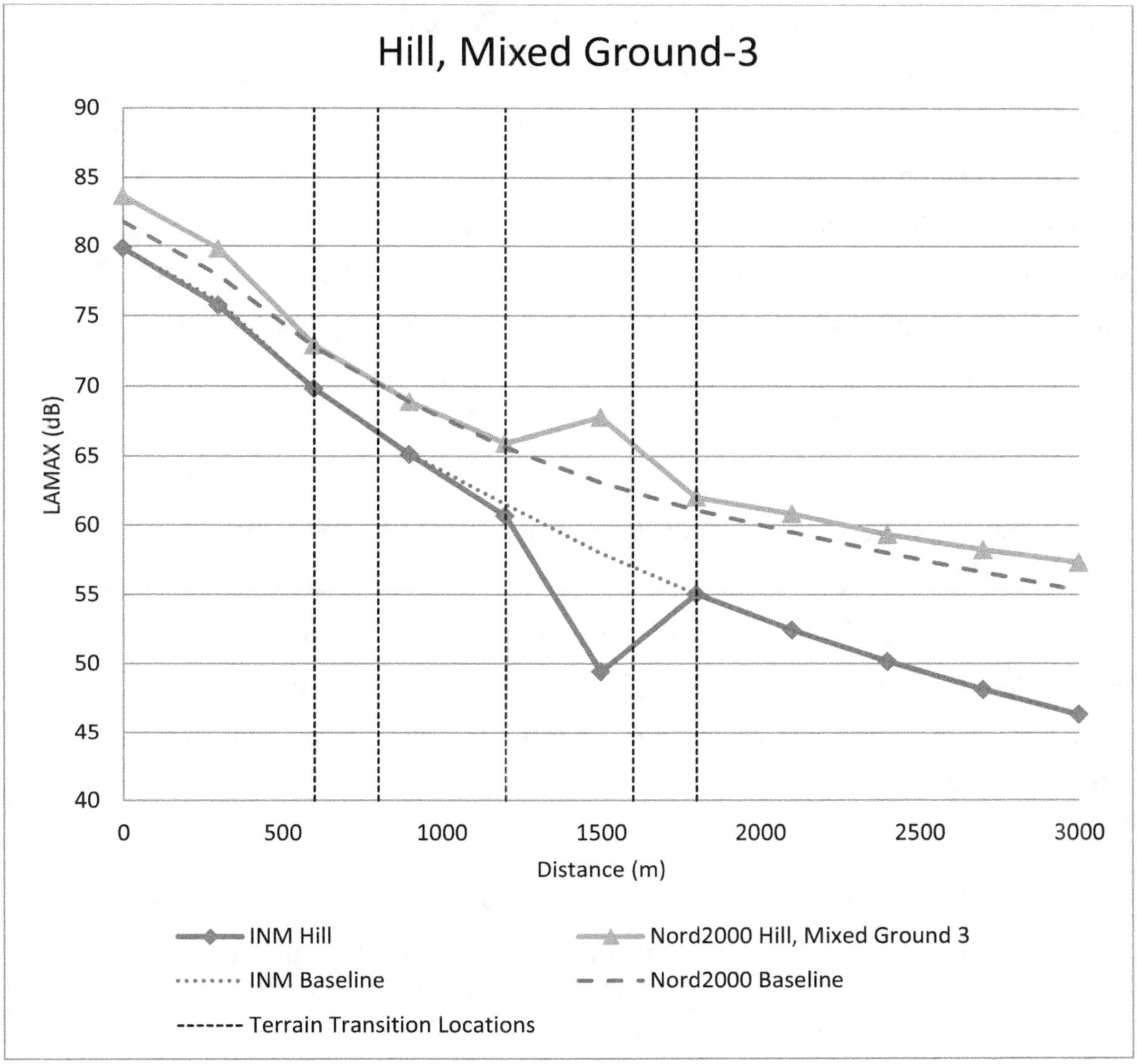

Hill, Mixed Ground-3

Figure 42. Hill Terrain, Mixed Ground Type-3 Results

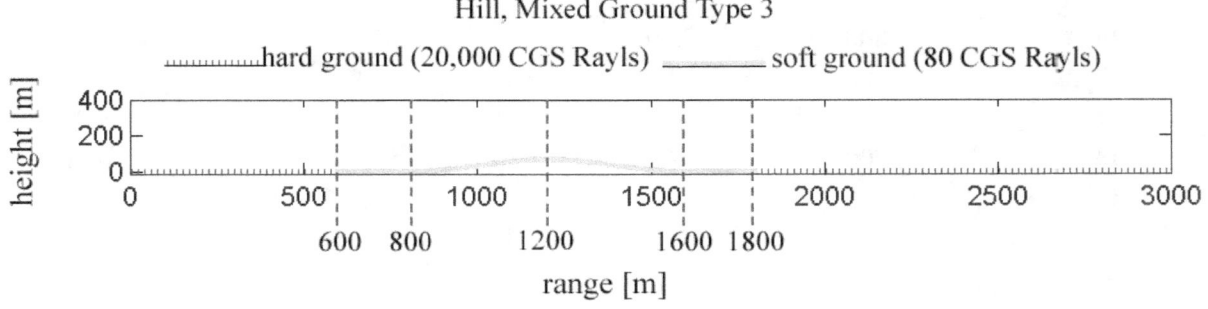

Figure 43. Hill Terrain, Mixed Ground Type-3 Comparison Terrain and Ground Type

Hill Terrain, Mixed Ground Type – 3, Turbulence Comparison

The results for the hill terrain, mixed ground type -3, turbulence comparison are shown in Table 51 and Figure 44. The LAMAX results are presented as the mean difference +/- the standard deviations of differences between INM and Nord2000 at the 11 receivers. The terrain and ground type are shown in Figure 45. The vertical dotted lines indicate ranges at which transitions in propagation conditions occur. Table 52 displays the difference between the Nord2000 hill terrain, mixed ground type-3 case and baseline results. Table 53 displays the difference between the hill terrain, mixed ground type-3, turbulence case and the hill terrain, mixed ground type-3 case.

Nord2000 computes LAMAX 7.7 +/- 4.5 dB greater than INM. The maximum difference between the two models for this comparison is 18.4 dB. The maximum difference occurs at receiver 14_5 which is located at approximately 1500 m. This is the first receiver to report sound levels past the peak of the hill. Table 53 highlights the effect of the turbulence on this comparison. The turbulence effect increases with distance, but only contributes up to 0.2 dB of the differences seen in this comparison.

Table 51. Hill Terrain, Mixed Ground Type-3, Turbulence Results

Receptor Row Fourteen	Distance (m)	INM (dB): Hill	Nord2000 (dB): Hill, Mixed ground 3, Turbulence	Difference (dB)
14_10	0	79.8	83.7	-3.8
14_9	300	75.7	79.8	-4.1
14_8	600	69.8	73.0	-3.1
14_7	900	65.1	68.9	-3.8
14_6	1200	60.7	66.0	-5.3
14_5	1500	49.4	67.8	-18.4
14_4	1800	55.0	62.1	-7.0
14_3	2100	52.4	60.9	-8.5
14_2	2400	50.2	59.4	-9.3
14_1	2700	48.1	58.4	-10.3
14_0	3000	46.3	57.5	-11.2

Table 52. Hill Terrain, Mixed Ground Type-3, Turbulence Comparison, Nord2000 Difference from Baseline

Distance (m)	Nord2000 (dB): Baseline	Nord2000 (dB): Hill, Mixed ground 3, Turbulence	Difference (dB)
0	81.8	83.7	1.9
300	77.9	79.8	1.9
600	72.8	73.0	0.1
900	68.8	68.9	0.0
1200	65.6	66.0	0.3
1500	63.1	67.8	4.7
1800	61.1	62.1	1.0
2100	59.5	60.9	1.4
2400	58.0	59.4	1.4
2700	56.6	58.4	1.8
3000	55.3	57.5	2.2

Table 53. Difference between Hill Terrain, Mixed Ground Type-3 Case with and without Turbulence

Receptor Row Fourteen	Nord2000 (dB): Hill, Mixed ground 3, Turbulence	Nord2000(dB): Hill, Mixed ground 3	Difference (dB)
14_10	83.7	83.7	0.0
14_9	79.8	79.8	0.0
14_8	73.0	72.9	0.1
14_7	68.9	68.9	0.0
14_6	66.0	65.9	0.1
14_5	67.8	67.8	0.0
14_4	62.1	62.0	0.1
14_3	60.9	60.8	0.1
14_2	59.4	59.3	0.1
14_1	58.4	58.2	0.2
14_0	57.5	57.3	0.2

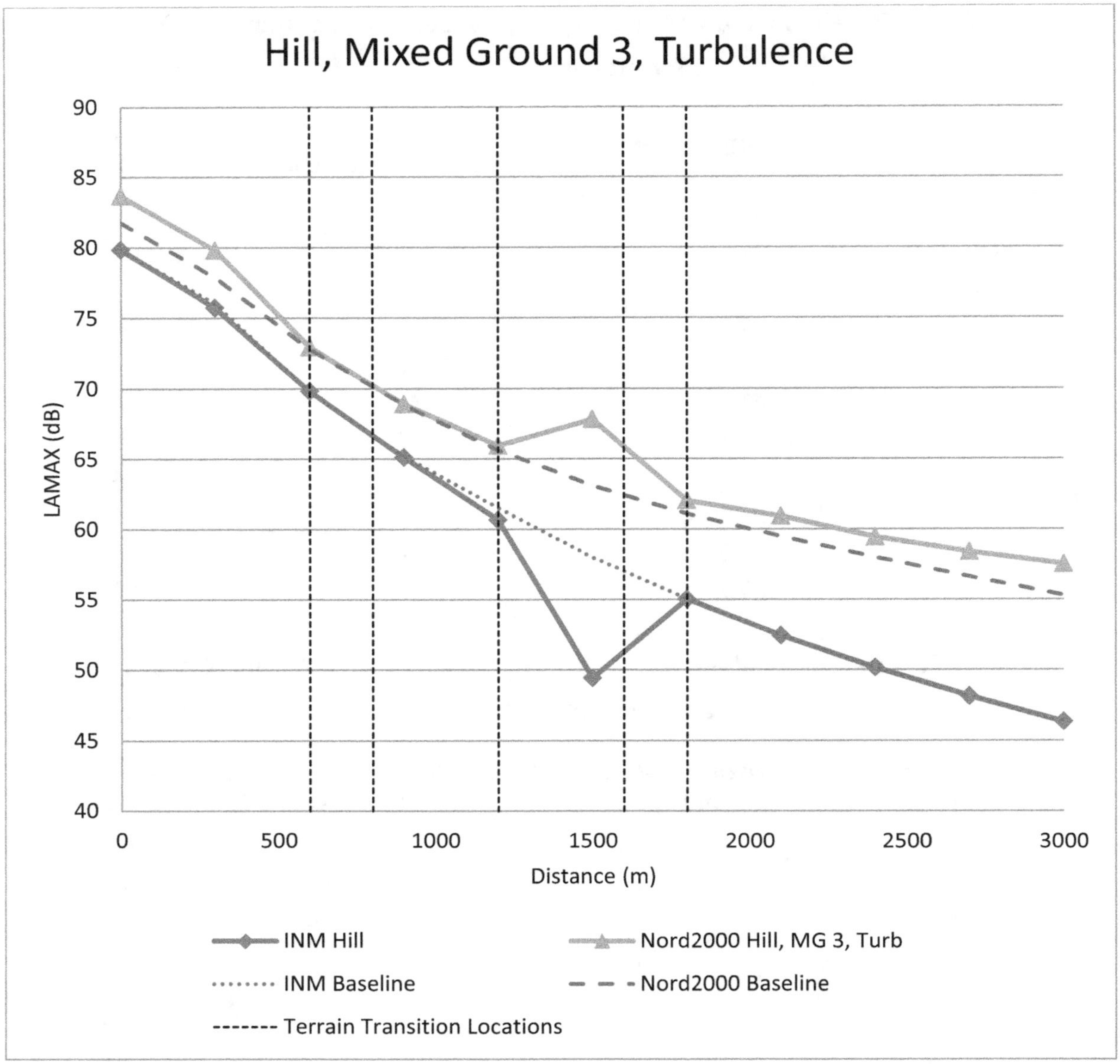

Figure 44. Hill Terrain, Mixed Ground Type – 3, Turbulence Comparison Terrain and Ground Type

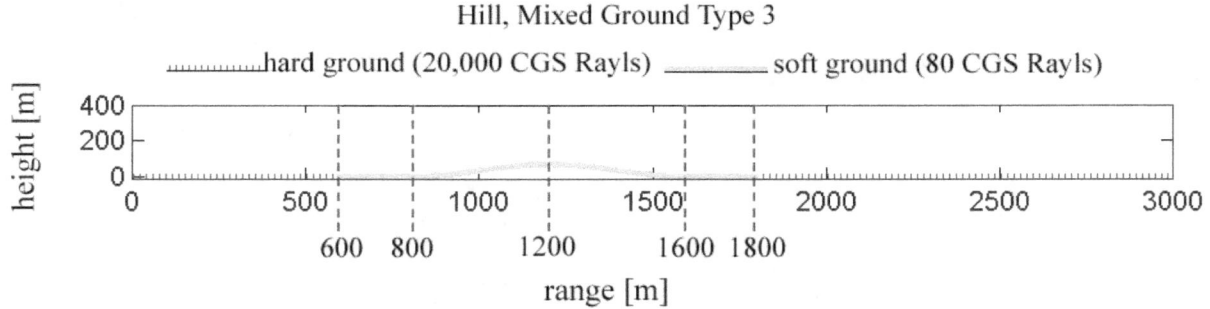

Figure 45. Hill Terrain, Mixed Ground Type-3

Hill Terrain, Positive Temperature Gradient

The results for the hill terrain, positive temperature gradient comparison are shown in Table 54 and Figure 46. The LAMAX results are presented as the mean difference +/- the standard deviations of differences between INM and Nord2000 at the 11 receivers. The terrain and ground type are shown in Figure 47. The vertical dotted lines indicate ranges at which transitions in propagation conditions occur. Table 52 displays the difference between the Nord2000 hill terrain, mixed ground type-3 case and baseline results. Table 53 displays the difference between the hill terrain, mixed ground type-3, turbulence case and the hill terrain, mixed ground type-3 case.

Nord2000 computes LAMAX 6.7 +/-3.9 dB greater than INM. The maximum difference between the two models for this comparison is 14.6 dB. The maximum difference occurs at receiver 14_5 which is located at approximately 1500 m. This is the first receiver to report sound levels past the peak of the hill. Table 56 and Table 57 show that the differences seen in the hill, positive temperature gradient comparison, are dominated by the positive temperature gradient. Individually, both the hill and positive temperature gradient comparisons show an increase in sound level, therefore it is not expected to see the effect of the combined comparison to result in a lower increase in sound level than in the individual comparisons. More investigation is needed to understand the interaction of the two effects to determine why the hill terrain feature contributes so little in this comparison.

Table 54. Hill Terrain, Positive Temperature Gradient Results

Receptor Row Fourteen	Distance (m)	INM (dB): Hill	Nord2000(dB): Hill, Positive Temp Gradient	Difference (dB)
14_10	0	79.8	81.8	-1.93
14_9	300	75.7	77.9	-2.1
14_8	600	69.8	73.0	-3.1
14_7	900	65.1	69.1	-4.0
14_6	1200	60.7	66.3	-5.7
14_5	1500	49.4	64.0	-14.6
14_4	1800	55.0	61.7	-6.7
14_3	2100	52.4	60.2	-7.7
14_2	2400	50.2	58.8	-8.7
14_1	2700	48.1	57.6	-9.5
14_0	3000	46.3	56.5	-10.1

Table 55. Hill Terrain, Positive Temp Gradient Comparison, Nord2000 Difference from Baseline

Distance (m)	Nord2000 (dB): Baseline	Nord2000(dB): Hill, Positive Temp Gradient	Difference (dB)
0	81.8	81.8	0.0
300	77.9	77.9	0.0
600	72.8	73.0	-0.1
900	68.8	69.1	-0.3
1200	65.6	66.3	-0.7
1500	63.1	64.0	-0.9
1800	61.1	61.7	-0.6
2100	59.5	60.2	-0.7
2400	58.0	58.8	-0.8
2700	56.6	57.6	-1.0
3000	55.3	56.5	-1.2

Table 56. Difference between Hill, Positive Temp Gradient Case and Positive Temp Gradient Case

Distance (m)	Nord2000(dB): Hill, Positive Temp Gradient	Nord2000 (dB): Positive Temp Gradient	Difference (dB)
0	81.8	81.8	0.0
300	77.9	77.9	0.0
600	73.0	73.0	0.0
900	69.1	69.1	0.0
1200	66.3	66.1	0.2
1500	64.0	63.7	0.3
1800	61.7	61.7	0.0
2100	60.2	60.2	0.0
2400	58.8	58.8	0.0
2700	57.6	57.6	0.0
3000	56.5	56.5	0.0

Table 57. Difference between Hill, Positive Temp Gradient Case and Hill Terrain Case

Distance (m)	Nord2000(dB): Hill, Positive Temp Gradient	Nord2000 (dB): Hill	Difference (dB)
0	81.8	81.8	0.0
300	77.9	77.9	0.0
600	73.0	72.8	0.2
900	69.1	68.9	0.2
1200	66.3	65.9	0.4
1500	64.0	65.0	-1.0
1800	61.7	61.4	0.3
2100	60.2	59.7	0.5
2400	58.8	58.1	0.7
2700	57.6	56.7	0.9
3000	56.5	55.3	1.2

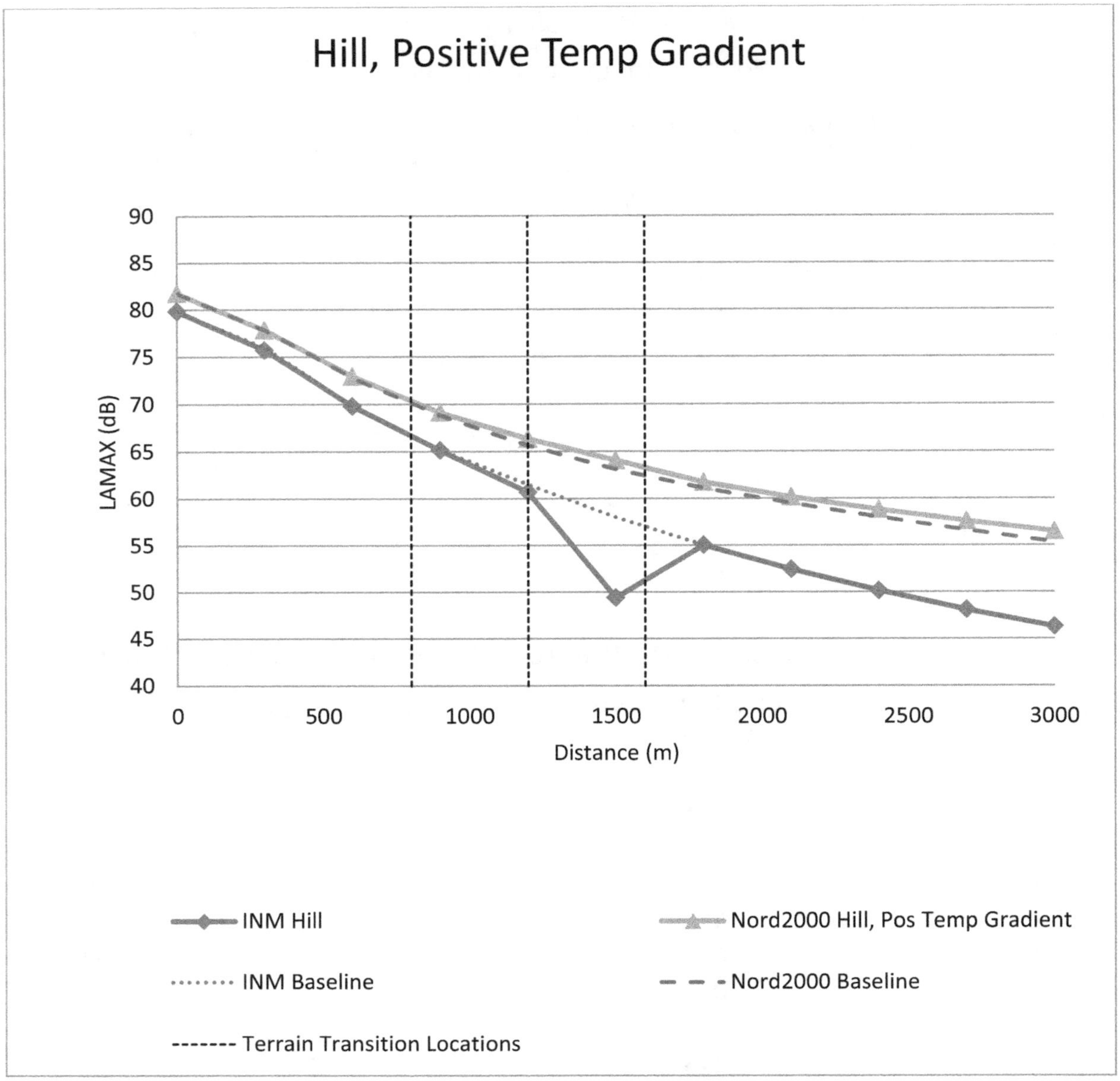

Hill, Positive Temp Gradient

Figure 46. Hill Terrain, Positive Temperature Gradient Results

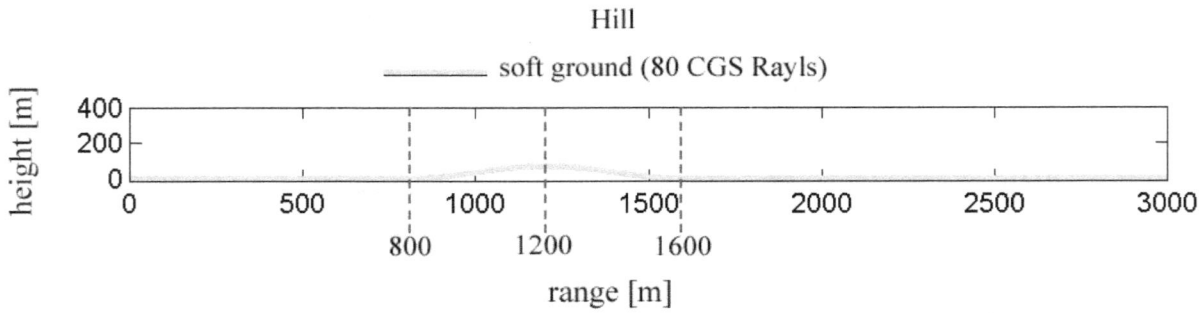

Figure 47. Hill Terrain and Ground Type

9.5 Hill Terrain, Multiple Parameter Variation Comparison Summary

All comparisons that include the hill terrain feature have unexpected trends in Nord2000. The hill terrain, mixed ground type -3 and the hill terrain, mixed ground type -3, with turbulence comparisons show an increase in sound level on the downslope portion of the hill. The hill, positive temperature gradient comparison shows almost no effect due to the hill terrain feature. Individually, both the hill and positive temperature gradient comparisons show an increase in sound level, therefore it is not expected to see the effect of the combined comparison to result in a lower increase in sound level than in the individual comparisons. Further investigation is needed to understand the unexpected trends in Nord2000 with the hill terrain feature present.

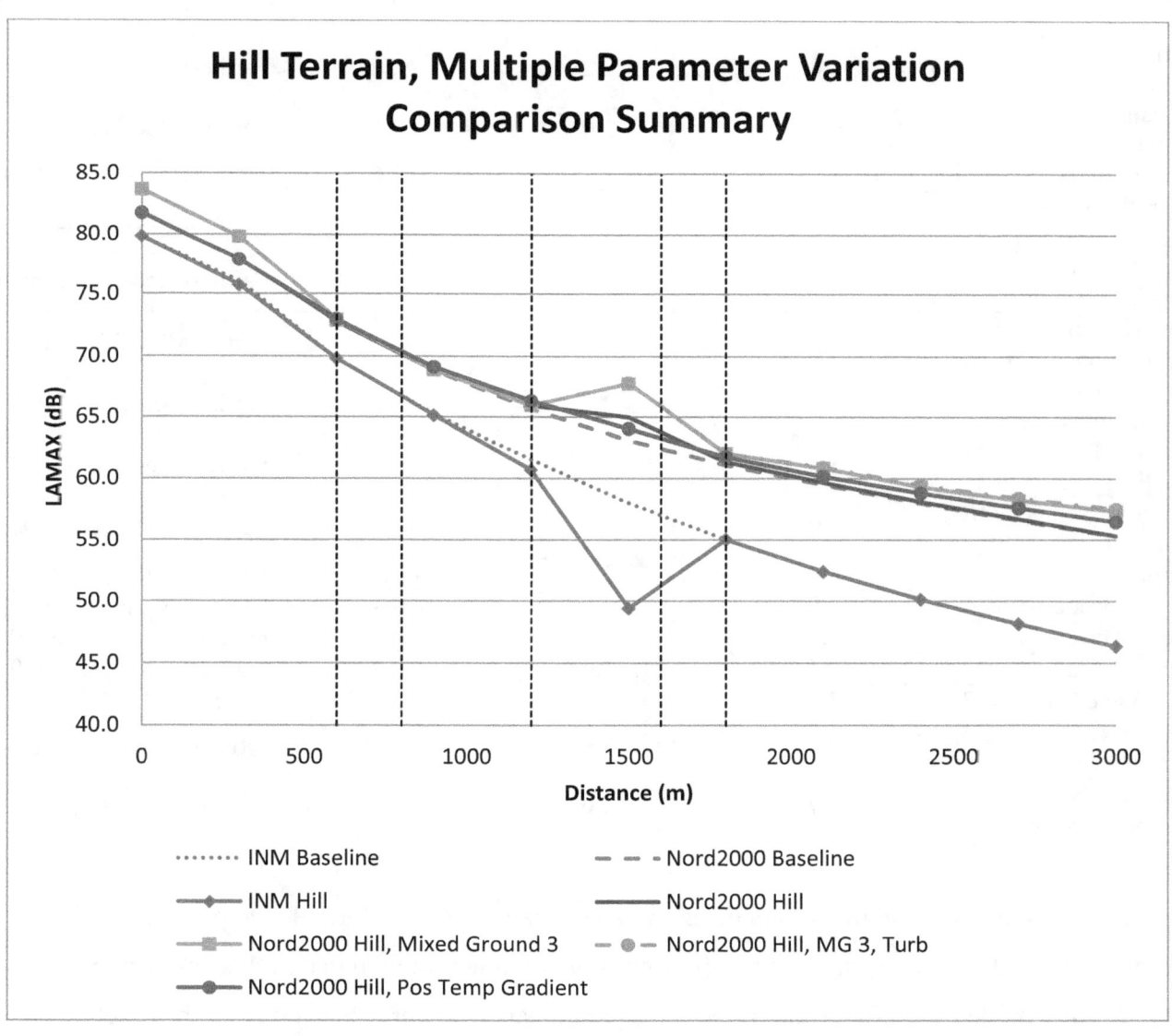

Figure 48. Hill Terrain, Multiple Parameter Variation Comparison Summary

10 RESULTS SUMMARY

The INM and Nord2000 comparison results from Sections 5 through 9 are summarized in Table 58. The average difference, standard deviation, maximum difference, and the location of the maximum difference between INM and Nord2000 are reported.

Table 58. INM, Nord2000 Comparison Results Summary

Comparison Description	Average Difference (dB) (INM-Nord2000)	Standard Deviation (dB)	Maximum difference (dB) (INM-Nord2000)	Location of Maximum Difference
Baseline	-5.3	+/- 2.6	9.0	Farthest Receiver (3000 m)
Effective Flow Resistivity	-7.0	+/- 2.4	-10.9	Farthest Receiver (3000 m)
Terrain- Downward Slope	-4.8	+/- 2.2	-8.0	Farthest Receiver (3000 m)
Terrain- Upward Slope	-6.1	+/- 3.0	-10.1	Farthest Receiver (3000 m)
Terrain- Hill	-6.4	+/- 4.0	-15.6	Receiver 14_5: ~1500 m
High Humidity, High Temperature	-5.4	+/- 2.9	-9.4	Farthest Receiver (3000 m)
High Humidity, Low Temperature	-5.4	+/- 3.0	-9.8	Farthest Receiver (3000 m)
Low Humidity, High Temperature	-4.5	+/- 2.1	-7.4	Farthest Receiver (3000 m)
Low Humidity, Low Temperature	-5.2	+/- 2.7	-8.5	Farthest Receiver (3000 m)
Positive Temperature Gradient	-5.8	+/- 3.1	-10.1	Farthest Receiver (3000 m)
Negative Temperature Gradient	-4.9	+/- 7.4	-12.8	Farthest Receiver (3000 m)
Turbulence	-5.3	+/- 2.6	-8.9	Farthest Receiver (3000 m)
Mixed Ground-1	-5.8	+/- 2.4	-10.4	Farthest Receiver (3000 m)
Mixed Ground-2	-6.6	+/- 3.0	-10.9	Farthest Receiver (3000 m)
Mixed Ground-3	-6.2	+/- 2.9	-10.9	Farthest Receiver (3000 m)
Hill, Mixed Ground-3	-7.6	+/- 4.5	-18.4	Receiver 14_5: ~1500 m
Hill, Mixed Ground-3, Turbulence	-7.7	+/- 4.5	-18.4	Receiver 14_5: ~1500 m
Hill, Positive Temperature Gradient	-6.7	+/- 3.9	-14.6	Receiver 14_5: ~1500 m

The largest average and maximum differences between INM and Nord2000 appear in the hill, mixed ground-3, with turbulence comparison. It is important to note that in this comparison, INM only models the hill terrain of the three parameters that were changed from the baseline. It is expected that differences between the two models would increase when multiple parameters not accounted for in INM are changed in Nord2000, however the Nord2000 hill terrain comparisons yield unexpected trends that require further investigation.

The INM and Nord2000 comparison results with respect to each model's respective baseline from Sections 5 through 9 are summarized in Table 59. The maximum difference is reported.

Table 59. Comparison Results Compared to Baseline Summary

Comparison Description	INM Maximum Difference from Baseline (dB) (baseline - comparison case)	Nord2000 Maximum Difference from Baseline (dB) (baseline - comparison case)
Effective Flow Resistivity	-	-2.0
Terrain- Downward Slope	-0.8	0.3
Terrain- Upward Slope	1.1	-0.4
Terrain- Hill	8.6	-1.9
High Humidity, High Temperature	2.5	2.1
High Humidity, Low Temperature	-1.3	-2.1
Low Humidity, High Temperature	2.5	4.1
Low Humidity, Low Temperature	5.6	6.1
Positive Temperature Gradient	-	-1.2
Negative Temperature Gradient	-	21.75
Turbulence	-	0.1
Mixed Ground-1	-	-1.9
Mixed Ground-2	-	-2.0
Mixed Ground-3	-	-2.0
Hill, Mixed Ground-3	-	4.7
Hill, Mixed Ground-3, Turbulence	-	4.7
Hill, Positive Temperature Gradient	-	-1.2

While the comparison of INM and Nord2000 results provides insight on how the two models differ from each other, when results from each model are compared to the respective baseline, the impact of the changing parameters are showcased. The comparison that highlights the parameter with the most impact when compared to its baseline is the negative temperature gradient comparison in Nord2000.

11 CONCLUSIONS AND RECOMMENDATIONS

11.1 Conclusions

General trends that appear throughout the comparisons described in this report show that (1) Nord2000 and INM have the best agreement at close distances and (2) Nord2000 computes higher LAMAX levels than INM with the exception of some (not all) receivers in the negative temperature gradient comparison.

While the following are based on the conditions explored in this report, some explanations for general causes of differences between INM and Nord2000 predictions are offered:

- Nord2000 and INM apply different atmospheric absorption parameters. This is one clear source of difference between Nord2000 and INM, causing Nord2000 to predict higher levels than INM does. These differences grow with range. However, the replacement of SAE-ARP-866A by SAE-ARP 5534 will reduce this effect in future releases of AEDT/INM.
- In conditions where there is soft ground under the source and standard temperature and relative humidity, Nord2000 returns levels approximately 2 dB larger than INM at the receiver beneath the source. The difference increases to approximately 4 dB when Nord2000 uses hard ground under the source. Because the difference in atmospheric absorption from different standards is small for a slant distance equal to the source height, this is likely due to different methods of including the effect of reflection of sound off the ground: The lateral attenuation adjustment in INM is zero directly beneath the source, assuming the effect of the ground-reflected sound is included in the source data. Nord2000, conversely, assumes the source input represents only the direct sound, and adds a ground-reflected component, raising the predicted level. Resolving this discrepancy could decrease INM-Nord2000 differences by 2 to 4 dB.

The impact of parameters that are not included in AEDT/INM range from 5.3 +/- 2.6 dB (from the turbulence comparison) to -7.0 +/- 2.4 dB (from the effective flow resistivity comparison) when modeled individually and increase to 6.7 +/- 3.9 dB (from the hill, positive temperature gradient comparison) to 7.7 +/- 4.5 dB (from the hill, mixed ground-3, turbulence comparison) in the simple combinations described in this report. It is clear that there are notable effects and that the parameters not included in INM should be considered for inclusion. Observations related to more complicated propagation effects include:

- Representation of a hard ground in Nord2000 raises levels by approximately 1 to 2 dB across the receiver range. The effect of different ground types cannot be captured in the current version of INM for jet aircraft noise.
- The largest deviations from the general INM-Nord2000 comparison trends to occur when a shadow zone is formed, as in the case of the negative temperature gradient and hill terrain comparisons.
- The effect that shows the largest deviation when compared to the baseline (of the same model) occurs in negative temperature gradient comparison.
- The effect of turbulence in raising levels has been seen to be substantial for longer range propagation where grazing angles of the ground-reflected ray are very shallow however; turbulence had only a small effect for any condition with the source-receiver geometries used in this analysis.
- Terrain features were seen to exacerbate differences between INM and Nord2000. Nord2000 exhibited unexpected results for terrain conditions, predicting a small increase in level just beyond the peak of hill terrain, where a decrease might have been expected.

Applicability to aircraft was another concern in this research. In its current state, the Nord2000 method does not include an aircraft source model, nor has it been validated beyond 400 m. For this research, an in-house method was created to integrate with INM's source data. Aircraft specific adjustments that exist in INM are not included in Nord2000. Modifications to the source model would also need to be made to accommodate metrics other than LAMAX in Nord2000. Further validation and analysis are needed to determine the accuracy of Nord2000 beyond 400m.

11.2 Recommendations

From the research described in this report and the confidence in the results yielded, it is recommended to pursue specific adjustments, and not implement the full Nord2000 method into AEDT/INM. Adjustments currently recommended for implementation are additional ground types and the SAE-ARP 5534 atmospheric absorption standard, once published. Implementation of these recommendations could improve AEDT noise propagation by 4-6 dB.

Additional ground types could be implemented as user defined parameters, as in Nord2000, or ground classes could be created based off of the categories presented in Table 4. Additional options could be added to the existing drop-down menu for ground type selection in AEDT/INM as it currently exists.

Other parameters should be considered for inclusions but require additional research before recommendations can be made with confidence, see Section 11.2.1.

11.2.1 Recommendations for Additional Research

Nord2000 effects that require additional research, but show potential for inclusion in AEDT/INM are vertical temperature gradients, pre-defined weather classes and turbulence. Terrain effects, hill terrain in particular, also require additional research before a recommendation can be made with confidence. Measurement validation is strongly suggested to determine which method, ray tracing or empirical, is more accurate when differences between the models arise.

With the largest deviation from the Nord2000 baseline, the negative vertical temperature gradient comparison has identified a significant area for improvement in AEDT/INM. Additional research is needed to determine the most appropriate way to implement into AEDT/INM since the effect of the vertical temperature gradient is calculated in Nord2000 using an approximated equivalent sound speed profile.

Implementation of weather classes in INM/AEDT based on typical weather conditions is recommended. Twenty-five weather classes have been established in the Nord2000 method, but not implemented in the version of the Nord2000 code obtained from the software exchange. The established weather classes are based on wind speed and atmospheric stability. Additional research is needed to determine the most appropriate way to implement into AEDT/INM, since the effect is calculated in Nord2000 using an approximated equivalent sound speed profile. A simplified version of the weather class concept could be explored with the existing parameters in AEDT/INM.

The effect of turbulence in shadow zones is expected to contribute more than what is shown in this research. Since turbulence is only applied when the line of sight is broken between the source and receiver, it is suggested to model a much larger terrain feature and increase and deliberately place receivers around where the line of sight is broken. Long range comparisons should also be conducted to further explore the proof of concept investigation described in 0.

Additional research in terrain effects is recommended, predominantly due to the large differences seen in the comparison results, but also because the Nord2000 methodology has only been validated out to 400 m for non-flat terrain. Finer resolution of receivers should be deliberately placed along the terrain feature transitions. Engine installation effects and the refraction-scattering components of the lateral attenuation adjustment included in INM should also be investigated for contribution to the differences seen in INM and Nord2000 results.

Researching all effects in an extended study scope is recommended. It is suggested to run additional comparisons in Nord2000, INM, and AEDT in order to compare features included in AEDT that are not available in INM. Comparisons including higher altitude overflights, full flight studies, and full airport studies are recommended. Additional multiple parameter variation comparisons should also be investigated.

In addition, this research could be used to further other FAA noise modeling research, such as supplementing the modeling of aircraft noise under complicated propagation conditions with the hybrid propagation model (HPM)[10,11]. As part of a future research effort, Nord2000 could be run with the same test cases used in the HPM sensitivity analyses, in order to (a) validate its performance against other propagation methods, and (b) determine if Nord2000 could provide an additional, supplemental propagation method for inclusion in the HPM. Alternatively, many of the adjustments recommended for inclusion in AEDT/INM (such as atmospheric absorption and ground absorption) could be implemented as an update to the ray-tracing propagation method already in the HPM.

Appendix A References

1. Boeker, et al.: <u>Integrated Noise Model (INM) Version 7.0 Technical Manual</u>, Report No. FAA-AEE-08-01, Washington, D.C.: Federal Aviation Administration, January 2008.

2. Eurasto: <u>Nord2000 for road traffic noise prediction. Weather classes and statistics,</u> VTT Research Report No. VTT-R-02530-06, July 2006.

3. Kragh, et al.: <u>Nordic Environmental Noise Prediction Methods, Nord2000 Summary Report. General Nordic Sound Propagation Model and Applications in Source-Related Predictions Methods</u>, DELTA Acoustics & Vibration Report AV 1719/01, December 2001.

4. Kragh, et al.: <u>User's Guide Nord2000 Road,</u> DELTA Acoustics & Vibration Report AV 1171/06, May 2006.

5. Occupational Safety and Health Administration: <u>Appendix I:A-3. Sound Propagation, Noise and Hearing Conservation, http://www.osha.gov/dts/osta/otm/noise/health_effects/soundpropagation.html</u>.

6. Page: <u>Simulation of DIA 737 Radar Track Departures Using Procedure Steps with INM,</u> Presentation at SAE A-21, December 2005.

7. Plovsing, Kragh: <u>Nord2000. Comprehensive Outdoor Sound Propagation Model. Part 1: Propagation in an Atmosphere without Significant Refraction</u>, DELTA Acoustics & Vibration Report AV 1849/00, December 2001.

8. Plovsing, Kragh: <u>Nord2000.Validation of the Propagation Model</u>, DELTA Acoustics & Vibration Report AV 1117/06, March 2006.

9. Plovsing: <u>Nord2000. Comprehensive Outdoor Sound Propagation Model. Part 2: Propagation in an Atmosphere with Refraction</u>, DELTA Acoustics & Vibration Report AV 1851/00, March 2006.

10. Rosenbaum, et al.: <u>Assessment of the Hybrid Propagation Model, Volume 1: Analysis of Noise Propagation Effects</u>, Report No. DOT/FAA/AEE/2012-03, Washington, D.C.: Federal Aviation Administration, August 2012.

11. Rosenbaum, et al.: <u>Assessment of the Hybrid Propagation Model, Volume 2: Comparison with the Integrated Noise Model</u>, Report No. DOT/FAA/AEE/2012-04, Washington, D.C.: Federal Aviation Administration, August 2012.

12. Salomons: <u>Computational Atmospheric Acoustics</u>, Kluwer Academic Publishers, 2001.

13. Society of Automotive Engineers, Committee A-21, Aircraft Noise, <u>Method for Predicting Lateral Attenuation of Airplane Noise</u>, Aerospace Information Report No. 5662, Warrendale, PA: Society of Automotive Engineers, Inc., April 2006.

Appendix B Source Model for Nord2000 Analysis

Nord 2000 is designed to accept sound power levels as source data input, however adjustments were made to the Nord2000 software to allow for sound pressure level input to be accepted for this analysis.

For accurate comparisons between Nord2000 and INM, an in-house tool was developed to simulate INM's source data adjustment process, in order to adjust the INM acoustic source data (NPDs and spectral classes) into source data for Nord2000. The process is as follows:

(1) Un-weighted spectral data at 1000 ft (spectral classes) are A-weighted.

(2) Because the spectral data serves only to provide the relationship between 1/3-octave band levels, rather than the correct absolute levels, a constant calibration correction is added to the bands that causes their logarithmic sum to reflect the appropriate NPD data for a source at 1000 ft.

(3) A-weighting is removed from the calibrated, 1000 ft spectrum.

(4) The source levels are corrected to a distance of 1m by removing spherical spreading and atmospheric absorption.

(5) A-weighting is reapplied for the propagation calculations.

Even with this source transformation, differences in source level were found due to differing source definition approaches. INM uses NPD curve data that are predefined at 10 distances from the source (200, 400, 630, 1000, 2000, 4000, 6300, 10000, 16000, and 25000 ft). It then extrapolates or interpolates from the two closest distances to determine the source level at any given distance. In contrast, the in-house tool that prepares source data for input to Nord2000 directly calculates the source level from the NPD data at 1000 ft, based on spherical spreading and atmospheric absorption, assuming a given spectral class. Differences between these two methods occur because the NPD curves are developed with the spectrum of the specific aircraft, whereas the in-house tool uses spectral class data, which are representative of a group of aircraft with similar spectral characteristics, to determine the appropriate atmospheric absorption corrections. To insure an accurate comparison of the noise propagation methodologies, a correction factor has been applied to Nord2000. The correction factor is computed as the difference between the source level calculated at a given distance (including spherical spreading and atmospheric absorption) and the corresponding source level calculated with the INM NPD interpolation method. With the correction, the source definition between INM and Nord2000 agrees within 0.0083 dB.

Appendix C Atmospheric Absorption Data

The results reported in the following tables are averaged over all INM spectral classes for each slant distance. The standard deviation is also reported. See Section 8.2.2 for the data in graphical form.

The current atmospheric absorption standard used in INM is SAE-ARP-866A. The current atmospheric absorption standard used in the Nord2000 methodology is ISO 9613-1. SAE-ARP-5534 is a pending replacement for SAE-ARP-866A.

Table 60. SAE-ARP-866A High Humidity Data

866A Slant Distance (m)	Temperature: 90 °F Relative Humidity: 90%		Temperature: 59 °F Relative Humidity: 90%		Temperature: 40 °F Relative Humidity: 90%	
	Average (dB)	Stnd Dev	Average (dB)	Stnd Dev	Average (dB)	Stnd Dev
30.48	84.42	9.45	84.50	9.46	84.44	9.43
334.06	81.75	8.78	82.42	8.93	82.30	8.78
637.64	79.84	8.42	80.88	8.59	80.91	8.44
941.22	78.32	8.22	79.60	8.37	79.82	8.26
1244.80	77.03	8.13	78.51	8.23	78.89	8.16
1548.38	75.91	8.09	77.55	8.15	78.09	8.11
1851.96	74.91	8.09	76.69	8.11	77.36	8.08
2155.55	74.00	8.10	75.90	8.09	76.69	8.07
2459.13	73.17	8.11	75.17	8.09	76.07	8.07
2762.71	72.38	8.12	74.50	8.09	75.49	8.08
3066.29	71.65	8.14	73.87	8.10	74.94	8.09
3369.87	70.95	8.15	73.27	8.11	74.41	8.09
3673.45	70.29	8.16	72.70	8.12	73.91	8.10
3977.03	69.66	8.17	72.15	8.13	73.43	8.11
4280.61	69.05	8.18	71.63	8.14	72.96	8.12
4584.19	68.47	8.19	71.13	8.15	72.52	8.13
4887.77	67.91	8.19	70.65	8.15	72.08	8.14
5191.35	67.37	8.20	70.18	8.16	71.66	8.14
5494.93	66.84	8.21	69.73	8.17	71.26	8.15
5798.52	66.34	8.22	69.29	8.17	70.86	8.16
6102.10	65.84	8.23	68.86	8.18	70.48	8.16
6405.68	65.36	8.24	68.45	8.19	70.10	8.17
6709.26	64.89	8.25	68.05	8.19	69.74	8.17
7012.84	64.43	8.25	67.65	8.20	69.38	8.18
7316.42	63.99	8.26	67.27	8.20	69.03	8.18
7620.00	63.55	8.27	66.89	8.21	68.69	8.19

Table 61. ISO 9613-1 High Humidity Data

9613-1	Temperature: 90 °F Relative Humidity: 90%		Temperature: 59 °F Relative Humidity: 90%		Temperature: 40 °F Relative Humidity: 90%	
Slant Distance (m)	Average (dB)	Stnd Dev	Average (dB)	Stnd Dev	Average (dB)	Stnd Dev
30.48	84.42	9.45	84.49	9.45	84.40	9.41
334.06	81.78	8.79	82.58	8.95	82.43	8.77
637.64	79.96	8.41	81.20	8.65	81.21	8.45
941.22	78.60	8.22	80.03	8.45	80.25	8.23
1244.80	77.54	8.14	79.01	8.31	79.45	8.13
1548.38	76.69	8.11	78.09	8.21	78.74	8.12
1851.96	75.97	8.12	77.26	8.15	78.09	8.09
2155.55	75.36	8.13	76.50	8.11	77.50	8.08
2459.13	74.82	8.14	75.79	8.09	76.94	8.07
2762.71	74.33	8.14	75.14	8.08	76.42	8.07
3066.29	73.89	8.15	74.52	8.08	75.92	8.07
3369.87	73.48	8.16	73.95	8.08	75.45	8.07
3673.45	73.10	8.16	73.41	8.09	74.99	8.08
3977.03	72.74	8.16	72.90	8.10	74.55	8.09
4280.61	72.41	8.17	72.42	8.11	74.12	8.09
4584.19	72.09	8.17	71.97	8.12	73.71	8.10
4887.77	71.79	8.17	71.53	8.13	73.31	8.11
5191.35	71.50	8.17	71.12	8.14	72.93	8.12
5494.93	71.23	8.18	70.73	8.15	72.55	8.12
5798.52	70.96	8.18	70.35	8.16	72.19	8.13
6102.10	70.71	8.18	69.99	8.17	71.83	8.14
6405.68	70.47	8.19	69.64	8.18	71.48	8.14
6709.26	70.23	8.19	69.31	8.19	71.14	8.15
7012.84	70.01	8.20	68.99	8.20	70.81	8.16
7316.42	69.79	8.20	68.69	8.21	70.48	8.16
7620.00	69.58	8.21	68.39	8.22	70.17	8.17

Table 62. SAE-ARP-5534 High Humidity Data

5534	Temperature: 90 °F Relative Humidity: 90%		Temperature: 59 °F Relative Humidity: 90%		Temperature: 40 °F Relative Humidity: 90%	
Slant Distance (m)	Average (dB)	Stnd Dev	Average (dB)	Stnd Dev	Average (dB)	Stnd Dev
30.48	84.38	9.42	84.45	9.43	84.38	9.39
334.06	81.74	8.77	82.56	8.95	82.41	8.76
637.64	79.89	8.39	81.15	8.64	81.17	8.44
941.22	78.50	8.20	79.96	8.43	80.19	8.26
1244.80	77.43	8.13	78.91	8.29	79.37	8.16
1548.38	76.57	8.11	77.98	8.19	78.65	8.11
1851.96	75.86	8.12	77.13	8.14	78.00	8.08
2155.55	75.25	8.13	76.35	8.10	77.40	8.07
2459.13	74.70	8.14	75.63	8.08	76.83	8.07
2762.71	74.21	8.15	74.96	8.08	76.30	8.07
3066.29	73.77	8.15	74.34	8.08	75.79	8.07
3369.87	73.36	8.16	73.75	8.09	75.31	8.08
3673.45	72.97	8.16	73.21	8.09	74.84	8.08
3977.03	72.61	8.16	72.69	8.11	74.39	8.09
4280.61	72.27	8.17	72.20	8.12	73.95	8.10
4584.19	71.95	8.17	71.74	8.13	73.53	8.11
4887.77	71.65	8.17	71.30	8.14	73.12	8.12
5191.35	71.36	8.18	70.88	8.15	72.73	8.12
5494.93	71.08	8.18	70.48	8.16	72.34	8.13
5798.52	70.81	8.18	70.10	8.17	71.97	8.14
6102.10	70.56	8.19	69.74	8.18	71.60	8.14
6405.68	70.31	8.19	69.39	8.19	71.24	8.15
6709.26	70.08	8.20	69.05	8.20	70.90	8.16
7012.84	69.85	8.20	68.73	8.21	70.56	8.16
7316.42	69.63	8.21	68.42	8.22	70.22	8.17
7620.00	69.42	8.21	68.12	8.23	69.90	8.17

Appendix C
Atmospheric Absorption Data

Table 63. SAE-ARP-866A Standard Humidity Data

866A	Temperature: 90 °F Relative Humidity: 70%		Temperature: 59 °F Relative Humidity: 70%		Temperature: 40 °F Relative Humidity: 70%	
Slant Distance (m)	Average (dB)	Stnd Dev	Average (dB)	Stnd Dev	Average (dB)	Stnd Dev
30.48	84.42	9.45	84.48	9.45	84.36	9.40
334.06	81.75	8.78	82.33	8.88	81.93	8.63
637.64	79.84	8.42	80.80	8.54	80.47	8.30
941.22	78.32	8.22	79.55	8.33	79.37	8.16
1244.80	77.03	8.13	78.47	8.21	78.47	8.10
1548.38	75.91	8.09	77.52	8.14	77.68	8.08
1851.96	74.91	8.09	76.67	8.10	76.98	8.07
2155.55	74.00	8.10	75.89	8.09	76.34	8.08
2459.13	73.17	8.11	75.17	8.08	75.74	8.09
2762.71	72.38	8.12	74.50	8.09	75.18	8.10
3066.29	71.65	8.14	73.86	8.10	74.64	8.11
3369.87	70.95	8.15	73.27	8.11	74.14	8.12
3673.45	70.29	8.16	72.70	8.12	73.65	8.13
3977.03	69.66	8.17	72.15	8.13	73.19	8.14
4280.61	69.05	8.18	71.63	8.14	72.74	8.15
4584.19	68.47	8.19	71.13	8.15	72.31	8.15
4887.77	67.91	8.19	70.65	8.15	71.89	8.16
5191.35	67.37	8.20	70.18	8.16	71.48	8.16
5494.93	66.84	8.21	69.73	8.17	71.09	8.17
5798.52	66.34	8.22	69.29	8.17	70.71	8.17
6102.10	65.84	8.23	68.86	8.18	70.34	8.17
6405.68	65.36	8.24	68.45	8.19	69.97	8.18
6709.26	64.89	8.25	68.05	8.19	69.62	8.18
7012.84	64.43	8.25	67.65	8.20	69.27	8.18
7316.42	63.99	8.26	67.27	8.20	68.93	8.19
7620.00	63.55	8.27	66.89	8.21	68.60	8.19

Table 64. ISO 9613-1 Standard Humidity Data

9613-1	Temperature: 90 °F Relative Humidity: 70%		Temperature: 59 °F Relative Humidity: 70%		Temperature: 40 °F Relative Humidity: 70%	
Slant Distance (m)	Average (dB)	Stnd Dev	Average (dB)	Stnd Dev	Average (dB)	Stnd Dev
30.48	84.42	9.45	84.45	9.43	84.32	9.37
334.06	81.78	8.83	82.45	8.88	82.10	8.63
637.64	79.89	8.47	81.07	8.57	80.81	8.32
941.22	78.42	8.26	79.92	8.37	79.84	8.18
1244.80	77.25	8.16	78.92	8.24	79.04	8.11
1548.38	76.29	8.12	78.02	8.16	78.34	8.08
1851.96	75.48	8.12	77.20	8.12	77.71	8.07
2155.55	74.79	8.12	76.45	8.09	77.13	8.07
2459.13	74.18	8.13	75.75	8.07	76.58	8.08
2762.71	73.63	8.14	75.09	8.07	76.07	8.08
3066.29	73.14	8.15	74.47	8.07	75.58	8.09
3369.87	72.69	8.16	73.88	8.07	75.12	8.10
3673.45	72.27	8.16	73.33	8.08	74.67	8.11
3977.03	71.88	8.17	72.80	8.09	74.24	8.11
4280.61	71.51	8.17	72.29	8.10	73.82	8.12
4584.19	71.16	8.18	71.81	8.11	73.42	8.13
4887.77	70.83	8.18	71.35	8.12	73.03	8.14
5191.35	70.52	8.19	70.90	8.13	72.65	8.14
5494.93	70.22	8.19	70.48	8.15	72.29	8.15
5798.52	69.93	8.20	70.07	8.16	71.93	8.15
6102.10	69.66	8.20	69.67	8.17	71.58	8.16
6405.68	69.40	8.21	69.29	8.18	71.23	8.16
6709.26	69.14	8.22	68.93	8.19	70.90	8.17
7012.84	68.90	8.22	68.57	8.20	70.57	8.17
7316.42	68.66	8.23	68.23	8.21	70.25	8.17
7620.00	68.44	8.24	67.90	8.22	69.94	8.18

Table 65. SAE-ARP-5534 Standard Humidity Data

5534	Temperature: 90 °F Relative Humidity: 70%		Temperature: 59 °F Relative Humidity: 70%		Temperature: 40 °F Relative Humidity: 70%	
Slant Distance (m)	Average (dB)	Stnd Dev	Average (dB)	Stnd Dev	Average (dB)	Stnd Dev
30.48	84.38	9.43	84.42	9.42	84.30	9.36
334.06	81.74	8.82	82.43	8.87	82.07	8.62
637.64	79.81	8.45	81.01	8.55	80.76	8.31
941.22	78.32	8.24	79.85	8.35	79.78	8.17
1244.80	77.12	8.15	78.83	8.23	78.97	8.11
1548.38	76.15	8.12	77.91	8.15	78.26	8.08
1851.96	75.34	8.12	77.08	8.10	77.62	8.07
2155.55	74.65	8.13	76.31	8.08	77.03	8.07
2459.13	74.04	8.14	75.59	8.07	76.48	8.08
2762.71	73.49	8.15	74.92	8.07	75.95	8.09
3066.29	73.00	8.16	74.29	8.07	75.46	8.09
3369.87	72.54	8.16	73.69	8.08	74.98	8.10
3673.45	72.12	8.17	73.12	8.09	74.52	8.11
3977.03	71.73	8.17	72.58	8.10	74.08	8.12
4280.61	71.35	8.18	72.06	8.11	73.66	8.13
4584.19	71.00	8.18	71.57	8.12	73.25	8.14
4887.77	70.67	8.19	71.09	8.13	72.85	8.14
5191.35	70.36	8.19	70.64	8.15	72.46	8.15
5494.93	70.06	8.20	70.21	8.16	72.08	8.15
5798.52	69.77	8.20	69.79	8.17	71.72	8.16
6102.10	69.49	8.21	69.39	8.18	71.36	8.16
6405.68	69.23	8.22	69.00	8.19	71.01	8.17
6709.26	68.97	8.22	68.63	8.20	70.66	8.17
7012.84	68.73	8.23	68.27	8.21	70.33	8.18
7316.42	68.49	8.24	67.93	8.22	70.00	8.18
7620.00	68.26	8.24	67.59	8.23	69.68	8.18

Table 66. SAE-ARP-866A Intermediate Humidity Data

866A	Temperature: 90 °F Relative Humidity: 45%		Temperature: 59 °F Relative Humidity: 45%		Temperature: 40 °F Relative Humidity: 45%	
Slant Distance (m)	Average (dB)	Stnd Dev	Average (dB)	Stnd Dev	Average (dB)	Stnd Dev
30.48	84.42	9.44	84.37	9.41	84.18	9.34
334.06	81.75	8.78	81.91	8.67	81.04	8.39
637.64	79.84	8.42	80.36	8.34	79.38	8.14
941.22	78.32	8.22	79.16	8.18	78.21	8.08
1244.80	77.03	8.13	78.14	8.11	77.26	8.08
1548.38	75.91	8.09	77.26	8.08	76.45	8.10
1851.96	74.91	8.09	76.46	8.07	75.73	8.12
2155.55	74.00	8.10	75.72	8.08	75.08	8.13
2459.13	73.17	8.11	75.03	8.08	74.48	8.14
2762.71	72.38	8.12	74.39	8.09	73.92	8.15
3066.29	71.65	8.14	73.78	8.11	73.40	8.16
3369.87	70.95	8.15	73.19	8.12	72.91	8.16
3673.45	70.29	8.16	72.64	8.13	72.44	8.17
3977.03	69.66	8.17	72.10	8.14	71.99	8.17
4280.61	69.05	8.18	71.59	8.15	71.56	8.17
4584.19	68.47	8.19	71.10	8.15	71.15	8.17
4887.77	67.91	8.19	70.62	8.16	70.75	8.17
5191.35	67.37	8.20	70.16	8.17	70.37	8.18
5494.93	66.84	8.21	69.71	8.17	69.99	8.18
5798.52	66.34	8.22	69.28	8.18	69.63	8.18
6102.10	65.84	8.23	68.85	8.18	69.28	8.18
6405.68	65.36	8.24	68.44	8.19	68.94	8.19
6709.26	64.89	8.25	68.04	8.19	68.60	8.19
7012.84	64.43	8.25	67.65	8.20	68.28	8.20
7316.42	63.99	8.26	67.26	8.21	67.96	8.20
7620.00	63.55	8.27	66.89	8.21	67.65	8.21

Table 67. ISO 9613-1 Intermediate Humidity Data

9613-1	Temperature: 90 °F Relative Humidity: 45%		Temperature: 59 °F Relative Humidity: 45%		Temperature: 40 °F Relative Humidity: 45%	
Slant Distance (m)	Average (dB)	Stnd Dev	Average (dB)	Stnd Dev	Average (dB)	Stnd Dev
30.48	84.41	9.44	84.33	9.39	84.15	9.31
334.06	81.80	8.86	82.05	8.68	81.31	8.38
637.64	79.88	8.52	80.63	8.37	79.88	8.15
941.22	78.31	8.31	79.51	8.21	78.86	8.09
1244.80	76.99	8.19	78.56	8.13	78.04	8.08
1548.38	75.86	8.13	77.72	8.09	77.32	8.08
1851.96	74.89	8.11	76.95	8.07	76.67	8.10
2155.55	74.04	8.11	76.23	8.07	76.08	8.11
2459.13	73.29	8.12	75.57	8.07	75.54	8.12
2762.71	72.62	8.13	74.93	8.07	75.03	8.13
3066.29	72.01	8.15	74.33	8.08	74.54	8.14
3369.87	71.45	8.16	73.75	8.09	74.08	8.15
3673.45	70.94	8.17	73.20	8.10	73.65	8.16
3977.03	70.46	8.18	72.67	8.11	73.22	8.16
4280.61	70.02	8.19	72.16	8.12	72.82	8.16
4584.19	69.60	8.20	71.66	8.13	72.43	8.17
4887.77	69.21	8.21	71.19	8.14	72.05	8.17
5191.35	68.84	8.22	70.73	8.15	71.68	8.17
5494.93	68.48	8.23	70.28	8.16	71.33	8.17
5798.52	68.15	8.24	69.84	8.17	70.98	8.17
6102.10	67.82	8.25	69.42	8.17	70.64	8.17
6405.68	67.51	8.26	69.01	8.18	70.31	8.18
6709.26	67.22	8.27	68.61	8.19	69.99	8.18
7012.84	66.93	8.28	68.22	8.20	69.67	8.18
7316.42	66.66	8.29	67.84	8.21	69.36	8.18
7620.00	66.39	8.30	67.47	8.21	69.06	8.18

Table 68. SAE-ARP-5534 Intermediate Humidity Data

5534	Temperature: 90 °F Relative Humidity: 45%		Temperature: 59 °F Relative Humidity: 45%		Temperature: 40 °F Relative Humidity: 45%	
Slant Distance (m)	Average (dB)	Stnd Dev	Average (dB)	Stnd Dev	Average (dB)	Stnd Dev
30.48	84.37	9.42	84.32	9.38	84.13	9.30
334.06	81.76	8.85	82.02	8.67	81.28	8.37
637.64	79.80	8.50	80.58	8.35	79.83	8.14
941.22	78.19	8.29	79.44	8.19	78.80	8.08
1244.80	76.85	8.17	78.47	8.12	77.96	8.08
1548.38	75.70	8.12	77.62	8.08	77.23	8.09
1851.96	74.72	8.11	76.83	8.07	76.58	8.10
2155.55	73.86	8.11	76.11	8.07	75.98	8.12
2459.13	73.10	8.13	75.42	8.07	75.43	8.13
2762.71	72.43	8.14	74.78	8.08	74.91	8.14
3066.29	71.82	8.15	74.16	8.09	74.42	8.15
3369.87	71.26	8.17	73.57	8.10	73.95	8.15
3673.45	70.74	8.18	73.00	8.11	73.50	8.16
3977.03	70.27	8.19	72.46	8.12	73.07	8.16
4280.61	69.82	8.20	71.93	8.13	72.66	8.17
4584.19	69.40	8.21	71.43	8.14	72.26	8.17
4887.77	69.00	8.22	70.94	8.15	71.87	8.17
5191.35	68.63	8.23	70.46	8.16	71.50	8.17
5494.93	68.27	8.24	70.00	8.17	71.13	8.17
5798.52	67.93	8.25	69.56	8.17	70.78	8.17
6102.10	67.61	8.26	69.12	8.18	70.43	8.18
6405.68	67.30	8.27	68.70	8.19	70.09	8.18
6709.26	67.00	8.28	68.29	8.20	69.76	8.18
7012.84	66.71	8.29	67.89	8.21	69.44	8.18
7316.42	66.44	8.30	67.50	8.22	69.12	8.18
7620.00	66.17	8.31	67.12	8.22	68.81	8.18

Table 69. SAE-ARP-866A Low Humidity Data

866A	Temperature: 90 °F Relative Humidity: 20%		Temperature: 59 °F Relative Humidity: 20%		Temperature: 40 °F Relative Humidity: 20%	
Slant Distance (m)	Average (dB)	Stnd Dev	Average (dB)	Stnd Dev	Average (dB)	Stnd Dev
30.48	84.30	9.39	83.99	9.28	83.97	9.32
334.06	81.41	8.60	80.25	8.24	79.17	8.22
637.64	79.57	8.27	78.39	8.09	76.79	8.09
941.22	78.12	8.13	77.09	8.09	75.18	8.12
1244.80	76.90	8.08	76.03	8.11	73.90	8.15
1548.38	75.83	8.07	75.13	8.13	72.82	8.16
1851.96	74.86	8.08	74.33	8.15	71.87	8.17
2155.55	73.97	8.09	73.61	8.16	71.03	8.18
2459.13	73.15	8.11	72.95	8.16	70.27	8.19
2762.71	72.37	8.12	72.34	8.17	69.56	8.20
3066.29	71.64	8.14	71.76	8.17	68.91	8.21
3369.87	70.95	8.15	71.22	8.17	68.30	8.22
3673.45	70.29	8.16	70.70	8.17	67.73	8.24
3977.03	69.66	8.17	70.21	8.18	67.19	8.25
4280.61	69.05	8.18	69.74	8.18	66.68	8.27
4584.19	68.47	8.19	69.29	8.18	66.20	8.28
4887.77	67.91	8.19	68.85	8.19	65.73	8.30
5191.35	67.37	8.20	68.43	8.20	65.29	8.32
5494.93	66.84	8.21	68.03	8.20	64.86	8.33
5798.52	66.33	8.22	67.63	8.21	64.45	8.35
6102.10	65.84	8.23	67.25	8.22	64.05	8.36
6405.68	65.36	8.24	66.88	8.22	63.66	8.38
6709.26	64.89	8.25	66.52	8.23	63.29	8.40
7012.84	64.43	8.25	66.16	8.24	62.92	8.41
7316.42	63.99	8.26	65.82	8.25	62.57	8.43
7620.00	63.55	8.27	65.48	8.26	62.22	8.44

Table 70. ISO 9613-1 Low Humidity Data

9613-1	Temperature: 90 °F Relative Humidity: 20%		Temperature: 59 °F Relative Humidity: 20%		Temperature: 40 °F Relative Humidity: 20%	
Slant Distance (m)	Average (dB)	Stnd Dev	Average (dB)	Stnd Dev	Average (dB)	Stnd Dev
30.48	84.27	9.38	83.98	9.25	83.97	9.31
334.06	81.52	8.65	80.63	8.25	79.48	8.20
637.64	79.68	8.33	79.03	8.09	77.45	8.10
941.22	78.19	8.17	77.88	8.08	76.11	8.12
1244.80	76.90	8.10	76.93	8.09	75.06	8.15
1548.38	75.75	8.07	76.10	8.11	74.17	8.16
1851.96	74.70	8.07	75.35	8.13	73.39	8.16
2155.55	73.74	8.07	74.67	8.14	72.69	8.16
2459.13	72.84	8.08	74.03	8.15	72.06	8.17
2762.71	72.00	8.10	73.43	8.16	71.47	8.17
3066.29	71.22	8.12	72.87	8.16	70.92	8.17
3369.87	70.48	8.13	72.33	8.17	70.40	8.18
3673.45	69.78	8.15	71.81	8.17	69.92	8.18
3977.03	69.12	8.17	71.32	8.17	69.46	8.19
4280.61	68.49	8.19	70.84	8.17	69.02	8.20
4584.19	67.90	8.20	70.38	8.17	68.59	8.20
4887.77	67.33	8.22	69.93	8.18	68.19	8.21
5191.35	66.79	8.24	69.50	8.18	67.80	8.22
5494.93	66.27	8.25	69.07	8.18	67.43	8.23
5798.52	65.78	8.27	68.66	8.18	67.07	8.24
6102.10	65.30	8.28	68.26	8.19	66.71	8.25
6405.68	64.85	8.30	67.87	8.19	66.37	8.26
6709.26	64.41	8.32	67.48	8.20	66.04	8.27
7012.84	63.99	8.33	67.11	8.20	65.72	8.28
7316.42	63.58	8.35	66.74	8.21	65.41	8.29
7620.00	63.18	8.37	66.38	8.21	65.10	8.30

Table 71. SAE-ARP-5534 Low Humidity Data

5534	Temperature: 90 °F Relative Humidity: 20%		Temperature: 59 °F Relative Humidity: 20%		Temperature: 40 °F Relative Humidity: 20%	
Slant Distance (m)	Average (dB)	Stnd Dev	Average (dB)	Stnd Dev	Average (dB)	Stnd Dev
30.48	84.26	9.37	83.97	9.25	83.93	9.28
334.06	81.48	8.64	80.59	8.24	79.42	8.19
637.64	79.61	8.31	78.98	8.09	77.38	8.10
941.22	78.09	8.16	77.81	8.08	76.02	8.13
1244.80	76.77	8.09	76.84	8.10	74.96	8.15
1548.38	75.59	8.07	76.00	8.12	74.06	8.16
1851.96	74.52	8.06	75.24	8.13	73.27	8.16
2155.55	73.54	8.07	74.54	8.15	72.56	8.16
2459.13	72.62	8.09	73.89	8.16	71.92	8.17
2762.71	71.76	8.11	73.29	8.16	71.32	8.17
3066.29	70.96	8.13	72.71	8.17	70.76	8.17
3369.87	70.20	8.15	72.16	8.17	70.24	8.18
3673.45	69.49	8.16	71.63	8.17	69.74	8.18
3977.03	68.81	8.18	71.13	8.17	69.28	8.19
4280.61	68.17	8.20	70.64	8.17	68.83	8.20
4584.19	67.57	8.22	70.17	8.18	68.40	8.21
4887.77	66.99	8.23	69.71	8.18	67.99	8.22
5191.35	66.44	8.25	69.26	8.18	67.59	8.23
5494.93	65.91	8.27	68.83	8.18	67.21	8.24
5798.52	65.41	8.29	68.41	8.19	66.84	8.25
6102.10	64.93	8.30	67.99	8.19	66.49	8.26
6405.68	64.47	8.32	67.59	8.20	66.14	8.27
6709.26	64.02	8.34	67.19	8.20	65.80	8.28
7012.84	63.59	8.35	66.81	8.21	65.48	8.29
7316.42	63.18	8.37	66.43	8.21	65.16	8.30
7620.00	62.78	8.39	66.06	8.22	64.84	8.31

Appendix D Noise Model Capability Chart

Table 72. Noise Model Capability Chart

Capability	Category	INM	Nord2000
Noise Database Structure	source	NPD data as a function of power (P) + spectral classes (average 1/3-octave spectra from 50 Hz-10kHz) + directivity (helicopters only)	sound power level (in 1/3-octave bands) + directivity
Database Coverage	source	115 commercial; 110 military (from NOISEMAP); 28 turboprop/piston; 17 Helicopters	No aircraft database. Source models are separate from propagation model (Road and Rail only). Sound power levels in third octave bands are given for 3 categories of road vehicles and for 20 trains.
Database Development	source	Manufacturers continually adding/updating per SAE-AIR-1845	Data on traffic--can be supplied by road authorities (Default data given in the User Guide may be useful) Topographical info--obtainable form digital maps/geographical info systems (data can be imported into computational software) Weather statistics--available from national road authorities
Overlapping Time Histories- Simultaneous events	source	Research Version or external to model (Time compression for TAUD)	n/a
Aircraft Performance	source	SAE-AIR-1845	n/a
Noise Sources	source	NPDs, Spectral Class Data	point source (sound power level) or moving source by simulation approach
Directivity	source	Behind start/takeoff roll, lateral directivity, helicopter specific directivity	Source models are separate from propagation model--existing source models contain horizontal directivity.
Aircraft Bank Angle	source	Yes	n/a
One-Third Octave Band Coverage	source /receiver	Standardized 50 Hz to 10 kHz for all data	25Hz-10kHz (No max levels for Passing Groups)
One-Third Octave Band Effects	path	Evaluated at center frequencies	Evaluated at center frequencies
Mixed Ground Impedance	path	Hard or Soft; Research Version: Fresnel Zone based distance weighting	Delany and Bazley Impedance Model

Appendix D
Noise Model Capability Chart

The Analysis of Modeling Aircraft Noise with the Nord2000 Noise Model

Capability	Category	INM	Nord2000
Ground/Terrain Effects (Blockage)	path	FHWA (Maekawa - Kurze/Anderson)	ground effect based on geometrical ray theory
Terrain Data Format	path	3CD, DEM, GridFloat (may be user-defined)	chain of straight line segments
Reflections/Scattering	path	Scattering -A component of lateral attenuation adjustment as a function of elevation angle	Yes. Urban areas-scattering taken into account from buildings and ground surfaces. Vegetation- statistical scattering model (external to Nord2000) average SPL at a specified distance from the source (does not consider sound shadow behind object and increase in SPL in front of object) - *Currently unusable- in need of additional research*
Weather Data	path	average annual	Sound speed profile that changes with altitude Weather classes
Sound Propagation	path	spherical spreading	Curved ray tracing
Sound Propagation-Weather Effect	path	atmospheric absorption	atmospheric absorption, refraction (wind and temperature) modeled by curved sound rays
Atmospheric Absorption Standard	path	SAE-ARP-866A	ISO 9613-1 (predicts pure-tone attenuation)
Noise Descriptors	receiver	Standard (A-, C- and tone-corr): SEL, DNL, CNEL, LEQ, LAeq(Day), LAeq(Night), Lmax, (%)TA, Ddose; (%)TAUD. + user-defined versions of all metrics	(A weight) L_{Aeq}, L_{den}, L_{AFmax} (defined as the level exceed by the noise from 5% of the vehicles in the actual category)
Change in Exposure Noise Metric	receiver	Yes	No
Interpolation/Extrapolation	receiver	Yes-Consistent with NOISEMAP	No-Propagation Code
overall use/scope	other	Aircraft and airport noise modeling	Sound propagation models using algorithms (vs. empirical models). Applications have been directed toward road and rail traffic. (wind turbines and airborne aircraft- propagation up to "*several*" kilometers)
Source Code	other	Available for Researchers only	Available under licensing
Source Code Language	other	C++	Fortran-NordRoad, C#-NorCalc, VB-Macro

Appendix D
Noise Model Capability Chart

**The Analysis of Modeling Aircraft Noise
with the Nord2000 Noise Model**

Capability	Category	INM	Nord2000
Case History	other	800+ Users World-wide over 30 Years	Developed from 1996-2001- used primarily in Nordic countries
Validation	other	INM 6.1 was validated for the vicinity of an airport with 3.2 dB agreement +/- 2.0 dB.	VALIDATED 2006: Distances up to 400 m with +/- 1dB (moderate terrain, wind). Dist 600-1000 m for flat ground only, +/- 2dB. VALIDATED 2001: Distances up to 200 m with +/- 2dB of overall A-weighted SPLs "in most cases". Larger distances have not been validated. AMBITION: "*Good*" up to 1000 m, "*acceptable*" up to 3000 m

www.ingramcontent.com/pod-product-compliance
Lightning Source LLC
Chambersburg PA
CBHW080303180526

45167CB00006B/2646